核化工与核燃料工程系列

# 核化工专业英语阅读


主　编　高　杨　何明键　周　羽

副主编　刘婷婷　王靖阳　张　萌　徐　真



哈爾濱工程大學出版社
Harbin Engineering University Press

## 内 容 简 介

本书共24章,内容涉及原子核、放射性衰变、天然放射性元素、人工放射性元素、溶剂萃取与离子交换、核材料、乏燃料后处理、放射性废物处理、放射性废物处置、放射性示踪剂、放射性核素发生器等。本书涵盖了核化工的主要研究方向及应用领域,可帮助学生进行专业英语学习、增加专业词汇量、提高专业文献阅读能力,亦可与专业课教学内容相呼应,加深学生对专业基础理论知识的理解。

本书可作为主修核化工、放射化学、核技术及相关专业的本科生教材,也可供其他相关专业的教师和学生以及从事放射化学研究和应用的工作人员参考。

**图书在版编目(CIP)数据**

核化工专业英语阅读 / 高杨, 何明键, 周羽主编
. — 哈尔滨 : 哈尔滨工程大学出版社, 2024.3
ISBN 978-7-5661-4311-2

Ⅰ. ①核… Ⅱ. ①高… ②何… ③周… Ⅲ. ①核化学-英语-教材 Ⅳ. ①O615.5

中国国家版本馆 CIP 数据核字(2024)第 058958 号

**核化工专业英语阅读**
HEHUAGONG ZHUANYE YINGYU YUEDU

**选题策划** 石 岭
**责任编辑** 王丽华
**封面设计** 李海波

---

**出版发行** 哈尔滨工程大学出版社
**社　　址** 哈尔滨市南岗区南通大街 145 号
**邮政编码** 150001
**发行电话** 0451-82519328
**传　　真** 0451-82519699
**经　　销** 新华书店
**印　　刷** 哈尔滨市海德利商务印刷有限公司
**开　　本** 787 mm×1 092 mm 1/16
**印　　张** 13
**字　　数** 432 千字
**版　　次** 2024 年 3 月第 1 版
**印　　次** 2024 年 3 月第 1 次印刷
**书　　号** ISBN 978-7-5661-4311-2
**定　　价** 45.00 元
http://www.hrbeupress.com
E-mail:heupress@hrbeu.edu.cn

---

# 前　言

核工业中所有与化学工程相关的环节都属于核化工的范畴，其中最重要的部分当属核燃料制造、核燃料循环和放射性废物处理。正因如此，核化工方面的工业能力是核能可持续发展和核威慑力量的重要保障。总体来说，西方世界在核化工领域的科技实力依旧领先，“师夷长技以制夷”的思路应该长期坚持。因此，学好英语，尤其是学好核化工领域的英语，对于了解核化工领域的科技前沿动态、开展国际科研合作、学习先进科研思路、领悟关键技术核心，都极其重要。

英语能力主要分为听、说、读、写四个方面，而对于缺乏专业英语知识的学生来说，英语阅读是最易入门的，也是最常用到的。编者为了补充核化工领域英语学习资料的不足，从英语阅读出发，编写了《核化工专业英语阅读》一书，书中内容涵盖了核工业中与核化工有关的英语阅读所需的大部分词汇。本书以指导专业英语阅读为目标，可能会忽略专业知识的完整性和系统性。例如书中涉及的核化工方面的英语章节，主要是作为英语阅读学习的素材和实例，难免存在对核化工领域知识的“断章取义”。但是，作为核化工专业英语阅读的学习资料，编者认为书中的英语阅读素材覆盖范围广泛、内容丰富全面，可以有效提升读者的核化工专业英语阅读能力，是一本值得核化工领域高校、研究院所、企事业单位的初学者使用的专业英语阅读学习资料。

由于编者水平有限，书中难免存在纰漏和不足之处，恳请读者批评指正。

编　者

2023 年 12 月

# Contents

# Chapter 1 Atoms and Atomic Nucleus

## 1.1 Species of Atomic Nuclei

The most elementary concept is that matter is composed of atoms. The atom is the smallest unit when a chemical element can be divided into without losing its chemical properties and contains nuclei and electrons. The constituents of atomic nuclei are called nucleons. The two types of nucleons are protons and neutrons. For example, the oxygen atom of mass number 16 has a nucleus which consists of 8 protons and 8 neutrons; since neutrons have no charge but are very similar to protons in mass, the net nuclear charge is +8. There are 8 extranuclear electrons in the neutral atom of oxygen.

The mass number $A$ is the total number of nucleons. Thus

$$A = N + Z$$

where $Z$ is the number of protons (= the atomic number) and $N$ is the number of neutrons. The elemental identity and the chemical properties are determined by the atomic number.

An element may be composed of atoms that, while having the same number of protons in the nuclei, have different mass numbers, therefore, different numbers of neutrons. Neon, for example, have an atomic number of 10, which means that the number of protons in the nuclei of all neon atoms is 10; however, 90% of the neon atoms in nature have 10 neutrons present in their nuclei while 10% of the atoms have 12 neutrons. Such atoms of constant $Z$ but different $A$ are called isotopes. The heavy hydrogen isotopes $^{2}H$ and $^{3}H$ are used so often in nuclear science that they have been given special names and symbols, deuterium (D) and tritium (T), respectively.

The word "nuclide" is used to designate any specific nuclear species, e. g., $^{16}O$, $^{14}C$, and $^{12}C$ are nuclides. Nuclides may be divided into groupings with common mass numbers and common neutron numbers. Isotopes are nuclides with a common number of proton ($Z$), whereas isobar is the term used to designate nuclides with a common number of nucleons ($A$), i. e., the same mass number. Nuclei with the same number of neutrons ($N$) but different atomic numbers are termed isotones.

In some cases, a nucleus may exist for some time in one or more excited states and it is differentiated on this basis. Such nuclei that necessarily have the same atomic number and mass number are called isomers. $^{60m}Co$ and $^{60g}Co$ are isomers; the $^{60m}Co$ nuclide exists in a high energy (excited) state and decays spontaneously by emission of a $\gamma$-ray to the lowest energy, ground state, designated by $^{60g}Co$.

## 1.2 Sizes and Masses of Nuclei

The scattering experiments of Rutherford showed that the nucleus occupies a very small portion of the total volume of the atom. Roughly, the radii of nuclei vary from 1/10 000 to 1/100 000 of the radii of atoms. While the atomic sizes are of the order of 100 pm (1 pm = $10^{-12}$ m), the common unit of nuclear size is the femtometer (1 fm = $10^{-15}$ m).

Experiments designed to study the size of nuclei indicate that the volumes of nuclei ($V_n$) are directly proportional to the total number of nucleons present, i. e.

$$V_n \propto A$$

Since for a sphere $V \propto r^3$, where $r$ is the radius of the sphere, for a spherical nucleus $r^3 \propto A$, or $r \propto A^{1/3}$. Using $r_0$ as the proportionality constant

$$r = r_0 A^{1/3}$$

The implications of this is that the nucleus is composed of nucleons packed closely together with a constant density from the centre to the edge of the nucleus.

Two forces are present in the nucleus, namely electrostatic forces between charged particles and gravitational forces between any objects that have mass. If only the electrostatic and gravitational forces existed in the nucleus, then it would be impossible to have stable nuclei composed of protons and neutrons. The gravitational forces are much too weak to hold the nucleons together compared to the electrostatic repelling forces among the protons. Since stable atoms of neutrons and protons do exist, there must be another attractive force acting within the nucleus. This force is called the nuclear force. Nuclear force is a strong attractive force that is independent of charge. It acts equally only between pairs of neutrons, pairs of protons, or a neutron and a proton. The nuclear force has a very short range; it acts only over distances approximately equal to the diameter of the nucleus. The attractive nuclear force between all nucleons drops off with distance much faster than the repulsive electrostatic force between protons. In stable atoms, the attractive and repulsive forces in the nucleus balance. If the forces do not balance, the nucleus will emit radiation in an attempt to achieve a stable configuration.

The universal mass unit, abbreviated u, is defined as one-twelfth of the mass of the $^{12}C$ atom which has been defined to be exactly 12 u. The absolute mass of a $^{12}C$ atom is obtained by dividing the value of 12 by the Avogadro number ($N_A = 6.02 \times 10^{23}$). The value for the mass of a $^{12}C$ atom, the nucleus plus the 6 extranuclear electrons, is thus $1.992\ 648 \times 10^{-23}$ g. Atomic masses are expressed in units of u relative to the $^{12}C$ standard. This text uses M to indicate masses in units of u, and m in units of kilograms, $m = M/10^3 N_A$.

In nuclear science it has been found convenient to use the atomic masses rather than nuclear masses. The number of electrons is always balanced in a nuclear reaction, and the changes in the binding energy of the electrons in different atoms are insignificant within the degree of accuracy used in the mass calculation. Therefore, the difference in atomic masses of reactants and products in a nuclear reaction gives the difference in the masses of the nuclei involved.

## 1.3 Mass Defect and Binding Energy

The masses of nuclei (in u) are close to the mass number $A$. Using the mass $^{12}C$ as the basis, the hydrogen atom and the neutron do not have exact unit masses. We would expect that the mass $M_A$ of an atom with mass number $A$ would be given by the number of protons ($Z$) times the mass of the hydrogen atom ($M_H$) plus the number of neutrons ($N$) times the mass of the neutron ($M_n$), i. e.

$$M_A \approx ZM_H + NM_n$$

For deuterium with one neutron and one proton in the nucleus, we would then anticipate an atomic mass of

$$M_H + M_n = 1.007\ 825 + 1.008\ 665 = 2.016\ 490\ \text{u}$$

When the mass of the deuterium atom is measured, it is found to be 2. 014 102 u. The difference between the measured and calculated mass values, which in the case of deuterium equals −0. 002 388 u, is called the mass defect ($\Delta M_A$):

$$\Delta M_A = M_A - ZM_H - NM_n$$

The equivalence between mass and energy is expressed by the famous equation:

$$E = mc^2$$

where $c$, the speed of light, is $2.998\times10^8$ m/s. We can calculate that one atomic mass unit is equivalent to 931. 5 MeV, where MeV is a million electron volts.

$$E = mc^2 = 931.5\Delta M_A$$

The relationship of energy and mass would indicate that in the formation of deuterium by the combination of a proton and neutron, the mass defect of 0. 002 388 u would be observed as the liberation of an equivalent amount of energy, i. e., $931.5\times0.002\ 388 = 2.224$ MeV. Indeed, the emission of this amount of energy is observed when a proton captures a low energy is observed when a proton captures a low energy neutron to form $^2$H. As a matter of fact, in this particular case, the energy liberated in the formation of deuterium has been used in the reverse calculation to obtain the mass of the neutron since it is not possible to determine directly the mass of the free neutron. All stable nuclei are found to have negative $\Delta M_A$; thus the term "defect" is used.

The larger the heat of formation the more stable the molecule since the more energy is required to decompose the molecule into its component atoms. Similarly, the energy liberated in the formation of a nucleus from its component nucleons is a measure of the stability of that nucleus. This energy is known as the binding energy ($E_B$) and has the same significance in nuclear science as the heat of formation has in chemical thermodynamics. We have seen that the binding energy of deuterium is 2. 224 MeV.

A better indication of the relative stability of nuclei is obtained when the binding energy is divided by the total number of nucleons to give the average binding energy per nucleon, $E_B/A$. For $^2$H it is 1. 11 for the bond between the two nucleons. For most nuclei the values of $E_B/A$ vary in the rather narrow range of 5−8 MeV. To a first approximation, therefore, $E_B/A$ is relatively constant which means that the total nuclear binding energy is roughly proportional to the total number of nucleons in the nucleus.

Figure 1. 1 shows that the $E_B/A$ values increase with increasing mass number up to a maximum around mass number 60 and then decrease. Therefore, the nuclei with mass numbers in the region of 60, i. e., nickel, iron, etc., are the most stable.

If two nuclides can be caused to react so as to form a new nucleus whose $E_B/A$ value is larger

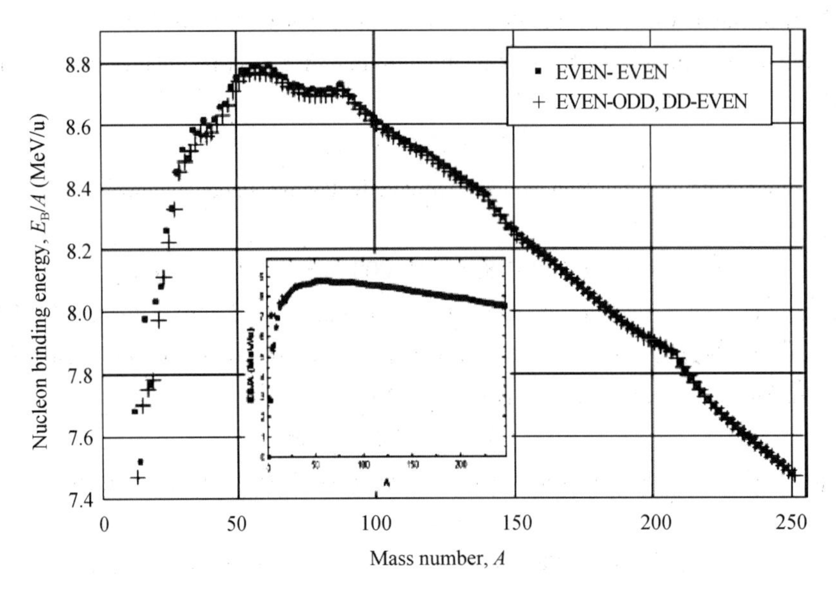


**Figure 1.1 Binding energy per nucleon ($E_B/A$) for the most stable isobars as function of mass number ($A$)**

than that of the reacting species, obviously a certain amount of binding energy would be released. The process which is called fusion is "exothermic" only for the nuclides of mass number below 60. A similar release of binding energy can be obtained if the elements with mass numbers greater than 60 are split into lighter nuclides with higher $E_B/A$ values. Such a process, whereby a nucleus is split into two smaller nuclides, is known as fission.

## 1.4 Neutron to Proton Ratio

If a graph is made (Figure 1.2) of the relation of the number of neutrons to the number of protons in the known stable nuclei, we find that in the light elements stability is achieved when the number of neutrons and protons are approximately equal ($N=Z$). However, with increasing atomic number of the element (i. e. along the $Z$-line), the ratio of neutrons to protons, the $N/Z$ ratio, for nuclear stability increases from unity to about 1.52 at lead. Thus pairing of the nucleons is not a sufficient criterion for stability: a certain ratio $N/Z$ must also exist. However, even this does not suffice for stability, because at high $Z$-values, a new mode of radioactive decay, α-emission, appears. Above bismuth the nuclides are all unstable to radioactive decay by α-particle emission, while some are unstable also to β-decay.

If a nucleus has an $N/Z$ ratio too high for stability, it is said to be neutron-rich. It will undergo radioactive decay in such a manner that the neutron-to-proton ratio decreases to approach more closely the stable value. In such a case the nucleus must decrease the value of $N$ and increase the value of $Z$, which can be done by conversion of a neutron to a proton. When such a conversion occurs within a nucleus, β-(or negatron) emission is the consequence, with the creation and emission of a negative β-particle.

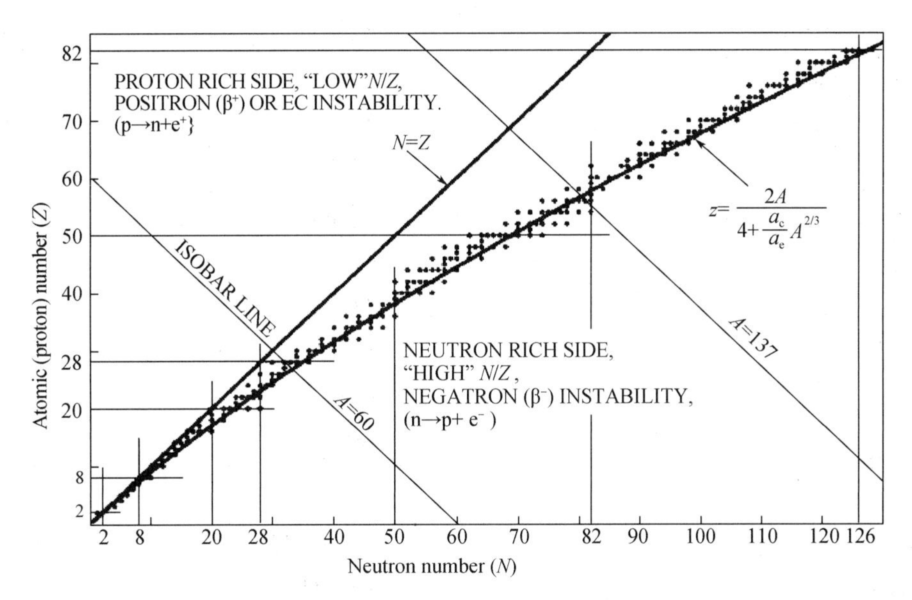


**Figure 1.2 Chart of stable nuclides as a function of their proton ($Z$) and neutron ($N$) numbers**

At extreme $N/Z$ ratios beyond the so-called neutron drip-line where the $N/Z$ ratio is too high for stability, or for highly excited nuclei, neutron emission is an alternative to β-decay.

If the β-ratio is too low for stability, then radioactive decay occurs in such a manner as to lower $Z$ and increase $N$ by conversion of a proton to neutron. This may be accomplished through positron emission, i. e., creation and emission of a positron, or by absorption by the nucleus of an orbital electron (electron capture, EC).

We can understand why the $N/Z$ ratio must increase with atomic number in order to have nuclear stability when we consider that the protons in the nucleus must experience a repulsive Coulomb force. The fact that stable nuclei exist means that there must be an attractive force tending to hold the neutrons and protons together. This attractive nuclear force must be sufficient in stable nuclei to overcome the disruptive Coulomb force. Conversely, in unstable nuclei there is a net imbalance between the attractive nuclear force and the disruptive Coulomb force. As the number of protons increases, the total repulsive Coulomb force must increase. Therefore, to provide sufficient attractive force for stability the number of neutrons increases more rapidly than that of the protons.

## 词汇

| | |
|---|---|
| atom | 原子 |
| nuclei | 原子核 |
| electron | 电子 |
| nucleon | 核子 |
| proton | 质子 |
| neutron | 中子 |

| | |
|---|---|
| extranuclear | 核外的 |
| mass number | 质量数 |
| atomic number | 原子序数 |
| neon atom | 氖原子 |
| isotope | 同位素 |
| hydrogen | 氢 |
| deuterium | 氘 |
| tritium | 氚 |
| nuclide | 核素 |
| isobar | 等压线 |
| designate | 命名 |
| isotone | 同中子异荷素 |
| isomer | 异构体 |
| decay | 衰变 |
| emission | 释放 |
| ground state | 基态 |
| radii | 半径 |
| spherical | 球形的 |
| proportionality | 比例性 |
| density | 密度 |
| electrostatic force | 静电力 |
| gravitational force | 引力 |
| repelling force | 排斥力 |
| attractive force | 吸引力 |
| nuclear force | 核力 |
| radiation | 辐射 |
| configuration | 构型 |
| mass unit | 质量单位 |
| abbreviate | 缩写 |
| Avogadro number | 阿伏伽德罗常量 |
| nuclear reaction | 核反应 |
| binding energy | 结合能 |
| reactant | 反应物 |
| anticipate | 预期 |
| mass defect | 质量亏损 |

| | |
|---|---|
| relationship of energy and mass | 质能关系 |
| liberation | 释放 |
| fusion | 核聚变 |
| exothermic | 放热的 |
| fission | 核裂变 |
| suffice | 足够 |
| lead | 铅 |
| radioactive decay | 放射性衰变 |
| bismuth | 铋 |
| neutron-rich | 丰中子 |
| conversion | 转换 |
| negatron | 负电子 |
| neutron drip-line | 中子滴线 |
| excited nuclei | 激发原子核 |
| positron | 正电子 |
| electron capture | 电子俘获 |
| repulsive Coulomb force | 库伦斥力 |

**注释**

1. The atom is the smallest unit when a chemical element can be divided into without losing its chemical properties and contains nuclei and electrons.

原子是一种化学元素在不失去化学性质的情况下可以被划分的最小单位，它包含原子核和电子。

2. An element may be composed of atoms that, while having the same number of protons in the nuclei, have different mass numbers, therefore, different numbers of neutrons.

一种元素可以由原子组成，这些原子在原子核中具有相同数量的质子，但具有不同的质量数，因此具有不同数量的中子。

3. Experiments designed to study the size of nuclei indicate that the volumes of nuclei ($V_n$) are directly proportional to the total number of nucleons present.

为研究核的大小而设计的实验表明，核的体积($V_n$)与存在的核子总数成正比。

4. The implications of this is that the nucleus is composed of nucleons packed closely together with a constant density from the centre to the edge of the nucleus.

这意味着原子核是由紧密堆积在一起的核子组成的，从原子核的中心到边缘密度恒定。

5. Two forces are present in the nucleus, namely electrostatic forces between charged particles and gravitational forces between any objects that have mass.

原子核中存在两种力，即带电粒子之间的静电力和任何有质量物体之间的引力。

6. The relationship of energy and mass would indicate that in the formation of deuterium by

the combination of a proton and neutron, the mass defect of 0.002 388 u would be observed as the liberation of an equivalent amount of energy.

质能关系表明,在质子和中子结合形成氘的过程中,可以观察到与0.002 388 u的质量亏损等量能量的释放。

7. As a matter of fact, in this particular case, the energy liberated in the formation of deuterium has been used in the reverse calculation to obtain the mass of the neutron since it is not possible to determine directly the mass of the free neutron.

事实上,在这种特殊情况下,由于不可能直接确定自由中子的质量,因此通过计算氘形成过程中释放的能量来反向获得中子的质量。

# Chapter 2 Radioactive Decay

## 2.1 Radioactive Decay

Radioactive decay is a spontaneous nuclear transformation that has been shown to be unaffected by pressure, temperature, chemical form, etc. This insensitivity to extranuclear conditions allows us to characterize radioactive nuclei by their decay period and their mode and energy of decay without regard to their physical and chemical condition.

The time dependence of radioactive decay is expressed in terms of the half-life ($T_{1/2}$), which is the time required for one-half of the radioactive atoms in a sample to undergo decay. In practice this is the time for the measured radioactive intensity (radioactivity of a sample) to decrease to one-half of its previous value. Half-lives vary from millions of years to fractions of seconds. The shortest half-life measurable today is about $10^{-18}$ s. Consequently, radioactive decay which occurs with a time period less than $10^{-18}$ s is considered to be instantaneous. At the other extreme, if the half-life of the radioactive decay exceeds $10^{15}$ a, the decay usually cannot be observed above the normal signal background present in the detectors. Therefore, nuclides which may have half-lives greater than $10^{15}$ a are normally considered to be stable to radioactive decay.

Radioactive decay involves a transition from a definite quantum state of the original nuclide to a definite quantum state of the product nuclide. The energy difference between the two quantum levels involved in the transition corresponds to the decay energy. This decay energy appears in the form of electromagnetic radiation and as the kinetic energy of the products.

The mode of radioactive decay is dependent upon the particular nuclide involved. Radioactive decay can be characterized by $\alpha$-, $\beta$-, and $\gamma$-radiation. Alpha decay is the emission of helium nuclide. Beta decay is the creation and emission of either electrons or positrons, or the process of electron capture. Gamma decay is the emission of electromagnetic radiation where the transition occurs between energy levels of the same nucleus. An additional mode of radioactive decay is that of internal conversion in which a nucleus loses its energy by interaction of the nuclear field with that of the orbital electrons, causing ionization of an electron instead of $\gamma$-ray emission. A mode of radioactive decay which is observed only in the heaviest nuclei is that of spontaneous fission in which the nucleus dissociates spontaneously into two roughly equal parts. This fission is accompanied by the emission of electromagnetic radiation and of neutrons.

## 2.2 Alpha Decay

Alpha decay is observed for the elements heavier than lead and for a few nuclei as light as the lanthanide elements. It can be written symbolically as

$$_{Z}^{A}X \longrightarrow _{Z-2}^{A-4}X + _{2}^{4}He \quad (2.1)$$

We use X to indicate any element defined by its nuclear charge, Z and Z−2 in this equation.

The decay energy can be calculated from the known atomic masses, because the binding energy released corresponds to a disappearance of mass. This energy is also called the $Q$-value of the reaction.

$$Q(MeV) = -931.5\Delta M(\mathrm{u}) \quad (2.2)$$

For α-decay we can define the $Q$-value as

$$Q_{\alpha} = -931.5(M_{Z-2} + M_{\mathrm{He}} - M_{Z}) \quad (2.3)$$

We always write the products minus the reactants within the parenthesis. A decrease in total mass in α-decay means a release of energy. The minus sign before the constant 931.5 is necessary to make $Q$ positive for spontaneous decay. For the decay reaction $^{238}U \rightarrow ^{234}Th + ^{4}He$, $Q_{\alpha} = -931.5$ (234.043 594+4.002 603−238.050 778 5) =4.274 MeV.

If the products are formed in their ground states, which is common for α-decay, the total decay energy is partitioned into the kinetic energies of the daughter nucleus ($E_{Z-2}$) and the helium nucleus ($E_{\alpha}$):

$$Q_{\alpha} = E_{Z-2} + E_{\alpha} \quad (2.4)$$

Because of conservation of energy and momentum,

$$E_{Z-2} = Q_{\alpha}M_{\alpha}/M_{Z} \quad (2.5)$$

and

$$E_{\alpha} = Q_{\alpha}E_{Z-2}/M_{Z} \quad (2.6)$$

From these equations, we can calculate the kinetic energy of the $^{234}Th$ daughter to be 0.072 MeV, while that of the α-particle is 4.202 MeV. Because of the large mass difference between the α-emitting nucleus and the helium atom, almost all of the energy is carried away with the α-particle.

Although the kinetic energy of the daughter nucleus is small in comparison with that of the α-particle, it is large (72 000 eV) in comparison with chemical binding energies (<5 eV). Thus the recoiling daughter easily breaks all chemical bonds by which it is bound to other atoms.

In 1904 it was observed that measurements on $^{218}Po$ (RaA), obtained from radon, led to contamination of the detection chamber by $^{214}Pb$ (RaB) and $^{214}Bi$ (RaC). This was explained by Rutherford as being due to daughter recoil in the α-decay of $^{218}Po$ in the sequence:

$^{222}Rn(\alpha,\ 3.8\ d)^{218}Po(\alpha,\ 3.05\ min)^{214}Pb(\beta^{-},\ 27\ min)^{214}Bi(\beta^{-},\ 20\ min)\ldots$

This recoil led to the ejection of $^{214}Pb$ into the wall of the instrument. The use of the recoil of the daughter to effect its separation was employed by O. Hahn beginning in 1909 and played a central role in elucidating the different natural radioactive decay chains.

The recoil may affect such chemical properties as the solubility or dissolution rate of compounds. For example, the dissolution of uranium from uranium rich minerals is considerably higher than one would expect from laboratory solubility data because α and U-atom recoil have moved U-atoms away from their normal sites in the mineral.

## 2.3 Beta Decay

The radioactive decay processes which are designed by the general name of β-decay include electron emission ($\beta^-$), positron emission ($\beta^+$), and electron capture (EC). If we use the β-decay of $^{137}Cs$ as an example, we can write

$$^{137}Cs \longrightarrow ^{137m}Ba + \beta^- \tag{2.7}$$

This β-decay must occur between discrete quantum levels of the parent nuclide $^{137}Cs$ and the daughter nuclide $^{137m}Ba$.

The quantum levels of nuclei are characterized by several quantum numbers, an important one being the nuclear spin. The spin value for the $^{137}Cs$ ground state level is 7/2, while that of $^{137m}Ba$ is 11/2. The electron emitted is an elementary particle of spin 1/2. In nuclear reactions the nuclear angular momentum must be conserved, which means that in radioactive decay processes the difference in total spin between reactant and products must be an integral value. The sum of the spin of the $^{137m}Ba$ and of the electron is 6, while that of the $^{137}Cs$ is 7/2. Therefore, the change in spin in the process would seem to be 5/2 spin units, which violates the rule for conservation.

The problem of "wrong" spin change led W. Pauli to the assumption that β-decay involves emission of still another particle which has been named the neutrino and given the symbol $\bar{\nu}$. The neutrino has a spin value of 1/2, an electric charge of 0, and a mass ≈ 0. It is therefore somewhat similar to the photon, which has neither mass, electric charge nor spin. However, while the photon readily interacts with matter, the neutrino does not. In fact the interaction is so unlikely that a neutrino has a very high probability of passing through the entire earth without reacting.

The spin attributed to the neutrino allows conservation of angular momentum. In our example, the total spin of the products would be 13/2, and when the spin of $^{137}Cs$, 7/2, is subtracted from this the result is 3 which is an acceptable integral value. Thus the decay reaction above is incomplete and must be written

$$^{137}Cs \longrightarrow ^{137m}Ba + \beta^- + \bar{\nu} \tag{2.8}$$

Notice we have replaced $\nu$ by $\bar{\nu}$, which is the designation of the antineutrino. Beta-decay theory has shown that antineutrinos $\bar{\nu}$ are emitted in electron decay, and "regular" neutrinos $\bar{\nu}$ in positron decay.

The energy released in β-decay is distributed between the neutrino, the electron and the recoil of the daughter nucleus. This latter will be much smaller than the first two and can be neglected in a first approximation. Therefore, the total β-decay energy can be considered to be distributed between the neutrino and the electron. For the decay $^{137}Cs \rightarrow ^{137m}Ba$ it can be shown that the total decay energy $Q_\beta$ is 0.514 MeV. This is also termed $E_{max}$. The neutrino energy spectrum is the complement of the β-particle energy spectrum. If the energy of the electron is 0.400 MeV, that of the neutrino is 0.114 MeV. If the electron energy is 0.114 MeV, the neutrino energy is 0.400 MeV.

In $\beta^-$-decay the average value of the $\beta^-$-particle energy is approximately 0.3 $E_{max}$. In positron emission, the average energy of the $\beta^+$-decay is approximately 0.4 $E_{max}$.

$\beta^-$-decay process can be written symbolically as follows:

$$^{A}_{Z}X \longrightarrow {}^{A}_{Z+1}X + {}^{0}_{-1}\beta + \bar{\nu} \tag{2.9}$$

However, if we take the electrons into account, the neutral parent atom has $Z$ orbital electrons, while the daughter atom, with a nuclear charge of $Z+1$, must capture an electron from the surroundings, in order to become neutral:

$$^{A}_{Z+1}X^{+} + e^{-} \longrightarrow {}^{A}_{z+1}X \tag{2.10}$$

Moreover, since the negatron emitted provides an electron to the surroundings, the total electron balance remains constant. As a result, in the calculation of the decay energy, it is not necessary to include the mass of the emitted $\beta$-particle as the use of the mass of the neutral daughter atom includes the extra electron mass. The equation for calculating the $Q$-value in negatron decay is thus:

$$Q_{\beta^-} = -931.5(M_{Z+1} - M_Z) \tag{2.11}$$

Positron decay can be written symbolically as

$$^{A}_{Z}X \longrightarrow {}^{A}_{Z-1}X^{-} + {}^{0}_{+1}\beta + \nu \longrightarrow {}^{A}_{Z-1}X + {}^{0}_{-1}e^{-} + {}^{0}_{+1}\beta + \nu \tag{2.12}$$

Here we must consider the net atomic charges. The daughter nucleus has an atomic number one less than the parent. This means that there will be one extra electron mass associated with the charge in atomic number. Moreover, an electron mass must also be included for the positive electron emitted. When $^{22}Na$ decays to $^{22}Ne$, there are 11 electrons included in the $^{22}Na$ atomic mass but only 10 in the $^{22}Ne$ atomic mass. Consequently, an extra electron mass must be added on the product side in addition to the electron mass associated with the positron particle. The calculation of the $Q$-value must therefore include two electron masses beyond that of the neutral atoms of the parent and the daughter.

$$Q_{\beta^+} = -931.5(M_{Z-1} + 2M_e - M_Z) \tag{2.13}$$

The EC decay process can be written symbolically as

$$^{A}_{Z}X \xrightarrow{EC} {}^{A}_{Z-1}X^{-} + \nu \tag{2.14}$$

The captured electron comes from one of the inner orbitals of the atom. Depending on the electron shell from which the electron originates, the process is sometimes referred to as K-capture, L-capture, etc. The probability for the capture of an electron from the K-shell is several times greater than that for the capture of an electron from the L-shell, since the wave function of K-electrons is substantially larger at the nucleus than that of L-electrons. Similarly, the probability of capture of electrons in higher order shells decreases with the quantum number of the electron shell.

The calculation of the decay energy in electron capture follows the equation

$$Q_{EC} = -931.5(M_{Z-1} - M_Z) \tag{2.15}$$

Note that like the case of the negatron decay, it is not necessary to add or subtract electron masses in the calculation of the $Q$-value in EC.

## 2.4 Gamma Emission and Internal Conversion

The α-and β-decay may leave the daughter nucleus in an excited state. This excitation energy is removed either by γ-ray emission or by a process called internal conversion.

The α-emission spectrum of $^{212}$Bi is shown in Figure 2.1. It is seen that the majority of the α-particles have an energy of 6.04 MeV, but a considerable fraction (about 30%) of the α-particles have higher or lower energies. This can be understood if we assume that the decay of parent $^{212}$Bi leads to excited levels of daughter $^{208}$Tl. This idea is supported by measurements showing the emission of γ-ray of energies which exactly correspond to the difference between the highest α-energy 6.08 MeV, and the lower ones.

Gamma rays produce very low density ionization in gases so they are not usually counted by ionization, proportional, or Geiger counters. However, the fluorescence produced in crystals such as sodium iodide make scintillation counting of γ-rays efficient. Gamma ray spectra can be measured with very high precision using semiconductor detectors.

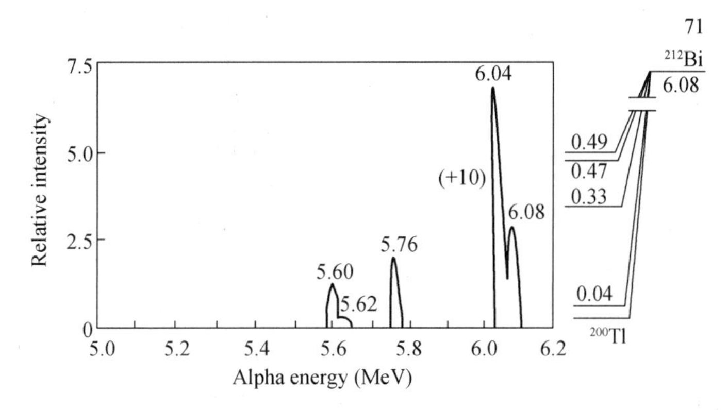


**Figure 2.1 α-emission spectrum from $^{212}$Bi→$^{208}$Tl. (According to E. B. Paul.)**

The decay energy in γ-emission is distributed between the γ-ray quantum ($E_\gamma$) and the kinetic energy of the recoiling product nucleus ($E_d$).

$$Q_\gamma = E_d + E_\gamma \tag{2.16}$$

The distribution of energy between the γ-ray and the recoiling daughter, according to

$$E_d = E_\gamma^2/(2m_d c^2) \tag{2.17}$$

shows that $E_d < 0.1\%$ of $E_\gamma$. The amount of kinetic energy of the recoiling nuclide is therefore so trivial that it may be neglected when only the γ-ray energy is considered.

A different process, called internal conversion, can occur within the atom undergoing radioactive decay. Because the wave function of an orbital electron may overlap that of the excited nucleus, the excitation energy of the nucleus can be transferred directly to the orbital electron, which escapes from the atom with a certain kinetic energy $E_e$. No γ-ray is emitted in an internal conversion process; it is an alternate mode to γ-ray emission of de-excitation of nuclei.

Internal conversion can be represented symbolically as

$$ {}^{Am}_{Z}X \longrightarrow {}^{A}_{Z}X^{+} + {}^{0}_{-1}e^{-} \longrightarrow {}^{A}_{Z}X \qquad (2.18) $$

Part of the nuclear excitation energy is required to overcome the binding energy $E_{be}$, of the electron in its electronic orbital. The remaining excitation energy is distributed between the recoiling daughter nucleus and the ejected electron $E_e$. The relationship is given by the equation

$$ Q_\gamma - E_{be} = E_d + E_e \qquad (2.19) $$

The ejected electron, known as the conversion electron, normally originates from an inner orbital, since its wave functions have greater overlap with the nucleus.

**词汇**

| | |
|---|---|
| spontaneous | 自发的 |
| instantaneous | 瞬间的 |
| quantum | 量子 |
| decay energy | 衰变能量 |
| electromagnetic | 电磁的 |
| fission | 裂变 |
| binding energy | 结合能 |
| reactant | 反应物 |
| recoiling | 反冲 |
| contamination | 污染 |
| nuclear spin | 核自旋 |
| angular momentum | 角动量 |
| neutrino | 中微子 |
| antineutrino | 反中微子 |
| energy spectrum | 能量谱 |
| excitation energy | 激发能 |
| internal conversion | 内转换 |
| ionization | 电离 |
| semiconductor detector | 半导体探测器 |
| conversion electron | 转换电子 |

**注释**

1. This insensitivity to extranuclear conditions allow us to characterize radioactive nuclei by their decay period and their mode and energy of decay without regard to their physical and chemical condition.

这种对核外条件的不敏感性使我们能够根据放射性原子核的衰变周期、衰变方式和能量来确定其特征,而无须考虑其物理和化学条件。

2. Radioactive decay involves a transition from a definite quantum state of the original nuclide to a definite quantum state of the product nuclide.

放射性衰变涉及从原始核素的确定量子态到产物核素的确定量子态的转变。

3. A mode of radioactive decay which is observed only in the heaviest nuclei is that of

spontaneous fission in which the nucleus dissociates spontaneously into two roughly equal parts.

只有在最重的原子核中才能观察到的一种放射性衰变模式是自发裂变,即原子核自发地解离成两个大致相等的部分。

4. Although the kinetic energy of the daughter nucleus is small in comparison with that of the α-particle, it is large (72 000 eV) in comparison with chemical binding energies (<5 eV).

虽然子核的动能与 α 粒子的动能相比很小,但与化学结合能(<5 eV)相比却很大(72 000 eV)。

5. The use of the recoil of the daughter to effect its separation was employed by O. Hahn beginning in 1909 and played a central role in elucidating the different natural radioactive decay chains.

O. Hahn 从 1909 年开始利用子代的反冲作用来实现其分离,这在阐明不同的天然放射性衰变链方面发挥了核心作用。

6. For example, the dissolution of uranium from uranium rich minerals is considerably higher than one would expect from laboratory solubility data because α and U-atom recoil have moved U-atoms away from their normal sites in the mineral.

例如,铀从富铀矿物中的溶解度远高于实验室预期的溶解度数据,因为 α 和铀原子反冲使铀原子偏离了它们在矿物中的正常位置。

7. In nuclear reactions the nuclear angular momentum must be conserved, which means that in radioactive decay processes the difference in total spin between reactant and products must be an integral value.

在核反应中,核角动量必须保持不变,这意味着在放射性衰变过程中,反应物和生成物之间的总自旋差必须是一个整数值。

8. In fact the interaction is so unlikely that a neutrino has a very high probability of passing through the entire earth without reacting.

事实上,这种相互作用的可能性非常小,中微子很有可能穿过整个地球而不发生反应。

9. However, if we take the electrons into account, the neutral parent atom has $Z$ orbital electrons, while the daughter atom, with a nuclear charge of $Z+1$, must capture an electron from the surroundings, in order to become neutral.

然而,如果我们考虑到电子,中性的母原子拥有 $Z$ 轨道电子,而核电荷为 $Z+1$ 的子原子则必须从周围环境中俘获一个电子才能成为中性原子。

10. The probability for the capture of an electron from the K-shell is several times greater than that for the capture of an electron from the L-shell, since the wave function of K-electrons is substantially larger at the nucleus than that of L-electrons.

从 K 层俘获电子的概率要比从 L 层俘获电子的概率大几倍,因为 K 层电子在原子核处的波函数要比 L 层电子的波函数大得多。

11. This idea is supported by measurements showing the emission of γ-ray of energies which exactly correspond to the difference between the highest α-energy 6.08 MeV, and the lower ones.

测量结果表明,发射的 γ 射线的能量正好与最高 α 能量 6.08 MeV 和较低 α 能量之间的差值相吻合。

12. Because the wave function of an orbital electron may overlap that of the excited nucleus,

the excitation energy of the nucleus can be transferred directly to the orbital electron, which escapes from the atom with a certain kinetic energy $E_e$.

由于轨道电子的波函数可能与受激原子核的波函数重叠,原子核的激发能量可以直接转移到轨道电子上,轨道电子以一定的动能 $E_e$ 逃离原子。

# Chapter 3 Nuclear Reactions

## 3.1 Conservation Laws in Nuclear Reactions

Reactions between an atomic nucleus and another particle are called nuclear reactions. In some such reactions, the nucleus is unchanged (elastic scattering), in others the original nucleus is excited to a higher energy state (inelastic scattering), in a third class new nuclei are formed (nuclear transmutations).

Let us begin by considering the nuclear reaction

$$^{4}_{2}He + ^{14}_{7}N \longrightarrow ^{17}_{8}O + ^{1}_{1}H \tag{3.1}$$

Most nuclear reactions are studied by inducing a collision between two nuclei where one of the reacting nuclei is at rest (the target nucleus) while the other nucleus (the projectile nucleus) is in motion. Such nuclear reactions can be described generically as:

projectile P + target T → emitted particle x and residual nucleus R

For example, the first reaction discussed above might occur by bombarding $^{14}N$ with α particles to generate an emitted particle, the proton and a residual nucleus $^{17}O$. A shorthand way to denote such reactions is, for the general case,

$$T(P, x)R$$

or for the specific example

$$^{14}N(\alpha, p)^{17}O$$

In a nuclear reaction, there is a conservation of the number of protons and neutrons (and thus the number of nucleons). Thus, the total number of neutrons (protons) on the left and right sides of the equations must be equal. There is also conservation of energy, linear momentum, angular momentum, and parity. Nuclear reactions, like chemical reactions, are always accompanied by a release or absorption of energy, and this is expressed by adding the term to the right hand side of the equation. Thus, a more complete statement of Eq. (3.1) is

$$^{4}_{2}He + ^{14}_{7}N \longrightarrow ^{17}_{8}O + ^{1}_{1}H + Q \tag{3.2}$$

The quantity $Q$ is the energy of the reaction or simply the reaction $Q$ value. Other than in chemistry, the convention is to assign positive $Q$ values to energy-releasing reactions (exoergic reactions) and negative $Q$ values to energy absorption (endoergic reactions). Another important difference between chemical reactions and nuclear reactions must be pointed out. In chemical reactions, macroscopic amounts of material undergo transmutation and heats of reaction are given per mole of the reactants. In nuclear reactions, single processes are considered and the $Q$ values are therefore given per nucleus transformed. For example, the reaction $^{14}N(\alpha, p)^{17}O$ has a $Q$ value of $-1.090$ MeV or $-4.56\times10^{-14}$ calories per $^{14}N$ atom transformed. To convert 1 mole of $^{14}N$ to $^{17}O$ would require an energy of $6.022\times10^{23}\times4.56\times10^{-14}$ cal $=2.75\times10^{10}$ cal. This is five orders of magnitude larger than the largest values observed for the heat of chemical reactions.

## 3.2 Reaction Types and Mechanisms

Nuclear reactions, like chemical reactions, can occur via different reaction mechanisms. Weisskopf has presented a simple conceptual model (Figure 3.1) for illustrating the relationships between the various nuclear reaction mechanisms.

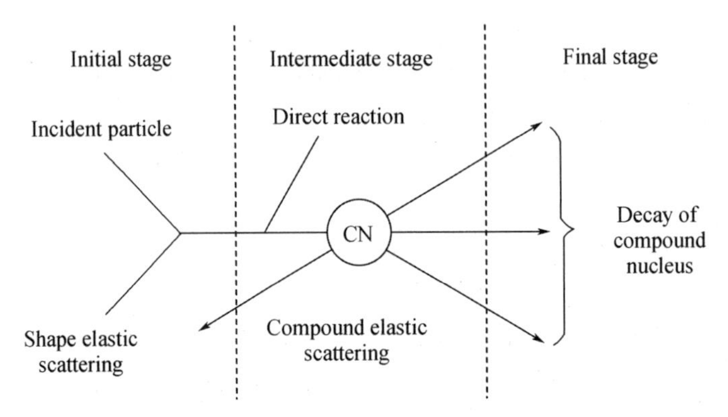


**Figure 3.1 Conceptual view of the stages of a nuclear reaction**

Consider a general nuclear reaction of the type A (a, b) B, bearing in mind that for some cases, the nuclei b and B may be identical to a and A. As the projectile a moves near the target nucleus A, it will have a certain probability of interacting with the nuclear force field of A, causing it to change direction but not to lose any energy ($Q=0$) (Figure 3.1). This reaction mechanism is called shape elastic scattering. If shape elastic scattering does not occur, then the projectile may interact with A via a two-body collision between the projectile and some nucleon of A, raising the nucleon of A to an unfilled level (Figure 3.1). If the struck nucleon leaves the nucleus, a direct reaction is said to have occurred. If the struck nucleon does not leave the nucleus, further two-body collisions may occur, and eventually the entire kinetic energy of the projectile nucleus may be distributed between the nucleons of the a+A combination leading to the formation of a compound nucleus C (Figure 3.1). Because of the complicated set of interactions leading to the formation of the compound nucleus, loosely speaking, it "forgets" its mode of formation, and its subsequent breakup only depends on the excitation energy, angular momentum of C, and so forth and not the nature of the projectile and target nuclei. Sometimes the compound nucleus may emit a particle of the same kind as the projectile (or even the projectile itself) with the same energy as the projectile had. If this happens, we say compound elastic scattering has occurred. In addition, C may decay into reaction products that are unlike the projectile or target nuclei.

## 3.3 Nuclear Reaction Cross Sections

The probability for a nuclear reaction to occur is expressed in terms of the reaction cross section. The geometric cross section that a nucleus presents to a beam of particles is $\pi r^2$. If we use $6\times10^{-15}$ m as an average value for the nuclear radius, the value of $\pi r^2$ becomes $3.14\times(6\times10^{-15})^2\approx10\times^{-28}$ $m^2$. This average geometric cross section of nuclei is reflected in the unit of reaction probability which is the barn, b, where 1 b = $10^{-28}$ $m^2$. The name barn has a historic background—cross sections of about $10^{-28}$ $m^2$ were said to be "big as a barn".

Consider the bombardment of a target containing $N_v$ atoms per $m^3$ by a homogeneous flux $\varphi_0$ of particles (Figure 3.2). The flux is expressed in units of particles $m^{-2}\cdot s^{-1}$. The target atoms $N_v$ refer only to the atoms of the species involved in the nuclear reaction. If a Li-Al alloy is bombarded to induce reactions with the lithium, $N_v$ is the number of lithium atoms per $m^3$ in the alloy, not the total of lithium and aluminum atoms. The change in the flux, $d\varphi=\varphi-\varphi_0$, may be infinitesimal as the particles pass through a thin section of target thickness $dx$. This change depends on the number of target atoms per unit area (i. e. $N_v dx$), the flux ($\varphi_0\approx\varphi$), and the reaction cross-section $\sigma$,

$$-d\varphi = \varphi\sigma N_v dx \tag{3.3}$$

The negative sign indicates that the flux decreases upon passing through the target due to the reaction of the particles with the target atoms, thus $-d\varphi$ is the number of reactions. Integration gives:

$$\varphi = \varphi_0 e^{-\sigma N_v x} \tag{3.4}$$

where $\varphi_0$ is the projectile flux striking the target surface. For targets which have a surface area of $S$ ($m^2$) exposed to the beam, for the irradiation time $t$, the total number of nuclear reactions $\Delta N$ is:

$$\Delta N = (\varphi_0 - \varphi)St \equiv \varphi_0 St(1 - e^{-\sigma N_v x}) \tag{3.5}$$

For a thin target in which the flux is not decreased appreciably upon passage through the target, i. e., $\sigma N_v x \ll 1$ and hence $e^{-\sigma N_v x}\approx 1-\sigma N_v x$, Eq. (3.4) can be reduced to:

$$\Delta N = \varphi_0 St\sigma N_v x = \varphi_0\sigma t N_v V = \varphi_0\sigma t N_t (\text{thin target}) \tag{3.6}$$

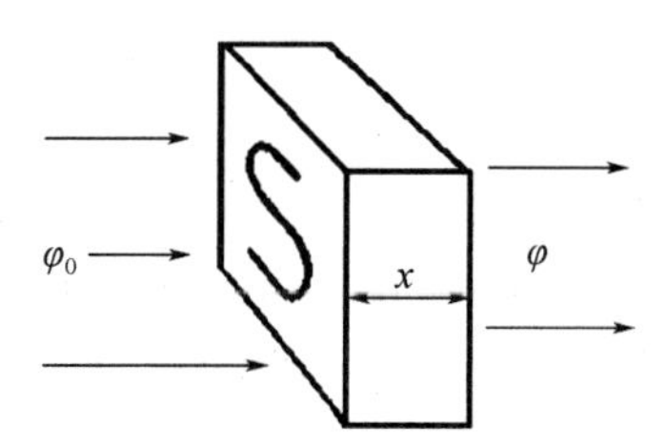


**Figure 3.2 Reduction of particle flux by absorption in a target**

where $V=S\cdot x$ is the target volume, and $N_t=N_v V$ is the number of target atoms. Notice that as a result of the product $S\cdot x$, which equals the volume of the target, the relationship on the right of Eq. (3.5) is independent of the geometry of the target and involves only the total number of atoms in it.

Eq. (3.5) can be used only when particle fluxes are homogeneous over the whole irradiated sample. In nuclear reactors, where the area of the sample is much smaller than the area of the flux, it is convenient to express the flux in terms of neutrons $m^{-2} \cdot s^{-1}$ and the target in terms of the total number of atoms, as above.

Frequently, the irradiated target consists of more than one nuclide which can capture bombarding particles to undergo reaction. The macroscopic cross section, which refers to the total decrease in the bombarding particle flux, reflects the absorption of particles by the different nuclides in proportion to their abundance in the target as well as to their individual reaction cross sections. Assuming that the target as a whole contains $N_v$ atoms $m^{-3}$ with individual abundances $\gamma_1$, $\gamma^2$, etc., for nuclides 1, 2, etc., the individual cross sections are $\sigma_1$, $\sigma_2$, etc. The macroscopic cross section $\hat{\sum}$ ($m^{-1}$) is

$$\hat{\sum} = N_v \sum_{1}^{n} \gamma_i \sigma_i \tag{3.7}$$

For a target which is x m thick one obtains

$$\varphi = \varphi_o e^{-\hat{\sum} x} \tag{3.8}$$

The value $\hat{\sum}^{-1}$ is the average distance a projectile travels between successive collisions with the target atoms (the mean free path).

## 3.4 Models for Nuclear Reactions

No single model is successful in explaining all the aspects of the various types of nuclear reactions. Here consider three models which have been proposed for explaining the results of nuclear reaction studies.

### The optical model

In the process of elastic scattering the direction of the particles is changed but none of the kinetic energy is converted to nuclear excitation energy. This would indicate that the reaction is independent of the internal structure of the nucleus and behaves much like the scattering of light from a crystal ball. Consequently, a model has been developed based on the mathematical techniques used in optics. Light shining on a transparent crystal ball is transmitted with some scattering and reflection but no absorption. Light shining on a black crystal ball is all absorbed and there is no transmission or scattering. In nuclear reactions the incoming particles are scattered in elastic scattering and are absorbed in induced transmutations. Therefore, if the nucleus is to act as a crystal ball it can be neither totally transparent nor totally black. The optical model of the nucleus is also known as the cloudy crystal ball model, indicating that nuclei both scatter and absorb the incoming particles.

The nucleus is described as a potential well containing neutrons and protons. The equation for the nuclear potential includes terms for absorption and scattering. This potential can be used to calculate the probability for scattering of incident particles and the angular distribution of the scattering. The model is in excellent agreement with experiments for scattering. Unfortunately, this model does not allow us to obtain much information about the consequences of the absorption of the particles which lead to inelastic scattering and transmutation.

### Liquid-drop model

N. Bohr offered a mechanism to explain nuclear reactions in nuclei which are excited into the continuum region. When a bombarding particle is absorbed by a nucleus, the kinetic energy of the bombarding particle plus the binding energy released by its capture provides the excitation energy of the compound nucleus. In this model, the compound nucleus becomes uniformly excited in a manner somewhat analogous to the warming of a small glass of water upon addition of a spoonful of boiling water. As the nucleons move about and collide in the nucleus, their individual kinetic energies vary with each collision just as those of molecules in a liquid change in molecular collisions. As this process continues, there is an increase in the probability that at least one nucleon will gain kinetic energy in excess of its binding energy (assuming the total excitation energy to be larger than the binding energy). That nucleon is then evaporated (i. e., leaves the nucleus) analogously to the evaporation of molecules from liquid surfaces.

The evaporation of the nucleon decreases the excitation energy of the residual nucleus by an amount corresponding to the binding energy plus the kinetic energy of the released nucleon. The evaporation process continues until the residual excitation energy is less than the binding energy of a nucleon. The excitation energy remaining at this point is removed from the nucleus by the emission of $\gamma$-rays.

### Direct interaction model

The compound nucleus model fails to explain some of the phenomena observed as the kinetic energy of the bombarding particle increases. Serber has suggested a mechanism that satisfactorily accounts for many features of nuclear reactions at bombardment energies above 50 MeV for protons, deuterons, and $\alpha$-particles. At such energies the relative speed between projectile and target nuclei is so high (near c) that the time available for the distribution of energy between all nucleons is too short and we can initially consider projectile and target nuclei as considering of fairly isolated nucleons. He proposed that high energy reactions occur in two stages.

(i) During the first stage, the nucleons in the incoming particle undergo direct collision with individual target nucleons. In these collisions the struck nucleon often receives energy much in excess of its binding energy. Consequently, after each collision both the nucleon belonging initially to the bombarding particle and the struck nucleon have some probability of escaping the nucleus since their kinetic energies are greater than their binding energies.

(ii) In the remaining nucleus the residual excitation energy is uniformly distributed. The reaction then enters its second and slower stage, during which the residual excitation energy is lost by nucleon evaporation. This stage resembles the compound nucleus process very closely.

### 词汇

nuclear reaction　　核反应

| | |
|---|---|
| elastic scattering | 弹性散射 |
| inelastic scattering | 非弹性散射 |
| transmutation | 嬗变 |
| nuclear transmutation | 核嬗变 |
| target nucleus | 靶核 |
| projectile nucleus | 射出核(弹核) |
| emitted particle | 发射的粒子 |
| residual nucleus | 剩余核 |
| bombarding | 碰撞 |
| energy | 能量 |
| linear momentum | 线性动量 |
| angular momentum | 角动量 |
| parity | 奇偶性 |
| energy-releasing | 能量释放 |
| exoergic reaction | 放能反应 |
| energy absorption | 能量吸收 |
| endoergic reaction | 吸能反应 |
| direct reaction | 直接反应 |
| compound nucleus | 复合核 |
| excitation energy | 激发能 |
| compound elastic scattering | 复合弹性散射 |
| cross section | 反应截面 |
| homogeneous flux | 均质通量 |
| lithium | 锂 |
| aluminum | 铝 |
| infinitesimal | 无穷小的 |
| crystal | 晶体 |
| transparent | 透明的 |
| transmit | 发射 |
| absorb | 吸收 |
| incoming particle | 入射粒子 |

**注释**

1. Most nuclear reactions are studied by inducing a collision between two nuclei where one of the reacting nuclei is at rest (the target nucleus) while the other nucleus (the projectile nucleus) is in motion.

大多数核反应是通过诱发两个核之间的碰撞来进行研究的,其中一个反应核处于静止状态(靶核),而另一个核(射出核)处于运动状态。

2. Nuclear reactions, like chemical reactions, are always accompanied by a release or absorption of energy, and this is expressed by adding the term to the right hand side of the equation.

像化学反应一样,核反应总是伴随着能量的释放或吸收,这可以通过在方程式的右边加上项(能量项)来表示。

3. As the projectile a moves near the target nucleus A, it will have a certain probability of interacting with the nuclear force field of A, causing it to change direction but not to lose any energy ($Q=0$).

当射出核 a 在靶核 A 附近移动时,它将有一定的概率与 A 的核力场相互作用,导致它改变方向,但不会失去任何能量($Q=0$)。

4. If the struck nucleon does not leave the nucleus, further two-body collisions may occur, and eventually the entire kinetic energy of the projectile nucleus may be distributed between the nucleons of the a+A combination leading to the formation of a compound nucleus C.

如果被击中的核子不离开原子核,可能会发生进一步双体碰撞,最终射出核的全部动能可能分布在 a+A 组合的核子之间,从而形成复合核子 C。

5. Because of the complicated set of interactions leading to the formation of the compound nucleus, loosely speaking, it "forgets" its mode of formation, and its subsequent breakup only depends on the excitation energy, angular momentum of C, and so forth and not the nature of the projectile and target nuclei.

由于导致复合核形成的一系列复杂的相互作用,宽泛地说,它(复合核)"忘记"了它的形成方式,它随后的解体只取决于激发能量、C 的角动量等,而不是射出核和靶核的性质。

6. The macroscopic cross section, which refers to the total decrease in the bombarding particle flux, reflects the absorption of particles by the different nuclides in proportion to their abundance in the target as well as to their individual reaction cross sections.

宏观截面是指轰击粒子通量的总减少量,反映了不同核素对粒子的吸收情况,以及各核素在靶核的丰度及其单反应截面的比例关系。

7. When a bombarding particle is absorbed by a nucleus, the kinetic energy of the bombarding particle plus the binding energy released by its capture provides the excitation energy of the compound nucleus.

当一个轰击粒子被一个原子核吸收时,轰击粒子的动能加上它被捕获所释放的结合能为复合核提供了激发能。

8. As this process continues, there is an increase in the probability that at least one nucleon will gain kinetic energy in excess of its binding energy (assuming the total excitation energy to be larger than the binding energy).

随着这一过程的继续,至少有一个核子获得超过其结合能的动能的概率会增加(假设总激发能大于结合能)。

# Chapter 4 Fission

## 4.1 The Fission Process

Fission has a unique importance among nuclear reactions. Apart from the nuclear reactions that drive the sun, no other nuclear reaction has had such a profound impact on the affairs of humans. The discovery of fission, and the developments that proceeded from it, have altered the world forever and have impinged on the consciousness of every literate human being.

Figure 4.1 shows a schematic view of the fission process. A nucleus with some equilibrium deformation absorbs energy, becomes excited, and deforms to a configuration known as the transition state or saddle point configuration. As it deforms, the nuclear Coulomb energy decreases (as the average distance between the nuclear protons increases) while the nuclear surface energy increases (as the nuclear surface area increases). At the saddle point, the rate of change of the Coulomb energy is equal to the rate of change of the nuclear surface energy. The formation and decay of this transition state nucleus is the rate-determining step in the fission process and corresponds to the passage over an activation energy barrier to the reaction. If the nucleus deforms beyond this point, it is irretrievably committed to fission. When this happens, then in a very short time the neck between the nascent fragments disappears and the nucleus divides into two fragments at the scission point. At the scission point, one has two highly charged, deformed fragments in contact with each other. The large Coulomb repulsion between the two fragments accelerates them to 90% of their final kinetic energy within $10^{-20}$ s. As these accelerated primary fragments move away from one another, they contract to more spherical shapes, converting their potential energy of deformation into internal excitation energy, that is, they get "hot". This excitation energy is removed by the emission of the "prompt" neutrons from the fully accelerated fragments and then, in competition with the last neutrons to be emitted, the nucleus emits gamma rays. Finally, on a longer time scale the neutron-rich fragments emit $\beta^-$ particles. Occasionally, one of these $\beta$ decays populates a high-lying excited state of a daughter that is unstable with respect to neutron emission, and the resulting emitted neutrons are called the "delayed" neutrons. Note that the energy release in fission is primarily in the form of the kinetic energies of the fragments not in the neutrons, photons, or other emitted particles. This energy is the "mass-energy" released in fission due to the increased stability of the fission fragments.

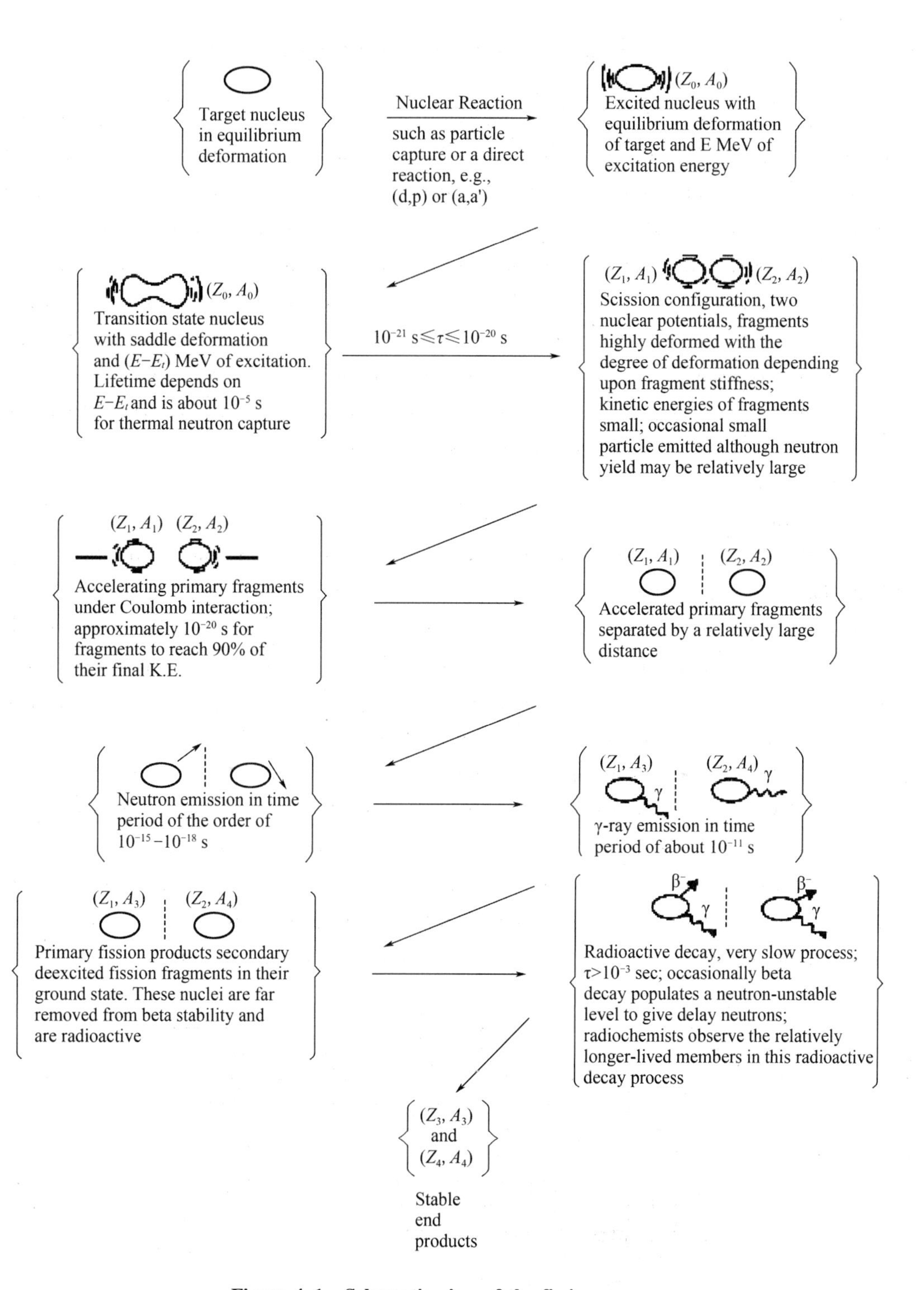


**Figure 4. 1 Schematic view of the fission process**

## 4.2 Spontaneous Fission

Nuclear fission can be either spontaneous or induced by particle (e.g., neutron) bombardment. In 1940, it was discovered that $^{238}U$ could decay by spontaneously fissioning into two large fragments (with a probability that was $5\times10^{-7}$ of that of undergoing α decay). Over 100 examples of this decay mode have been found since then. Spontaneous fission is a rare decay mode in the light actinides and increases in importance with increasing atomic number until it is a stability-limiting mode for nuclei with $Z\geqslant98$. The spontaneous fission half-lives change by a factor of $10^{29}$ in going from the longest lived uranium nuclei to the short-lived isotopes of fermium.

Spontaneous fission is a barrier penetration phenomenon similar to α or proton decay. The nucleus "tunnels" from its ground state through the fission barrier to the scission point. Therefore, we would expect the spontaneous fission (SF) half-life to have the form

$$t_{1/2}^{SF}=\frac{\ln 2}{fP} \tag{4.1}$$

where $f$ is the frequency of assaults on the fission barrier in the first minimum (about $10^{20}$/s) and $P$ is the barrier penetrability.

Since the discovery of the first spontaneously fissioning isomer, a number of other examples have been found. The positions of these nuclei in the chart of nuclides are shown in Figure 4.2. These isomers range from thorium to berkelium, forming an island with a point of maximum stability around $^{242}Am$. γ-ray decay back to the ground state limits the number of isomers with lower $Z$ and $N$ than those in this island, whereas spontaneous fission decay limits the number of cases with high $Z$ and $N$. The half-lives range from $10^{-9}$ to $10^{-3}$ s, whereas the ground-state half-lives are about $10^{25}-10^{30}$ times longer. The typical excitation energy of these isomers is 2–3 MeV. Spectroscopic studies of the transitions between the states in the second minimum have shown that the moments of inertia associated with the rotational bands are those expected for an object with an axes ratio of 2:1—a result confirmed in quadrupole moment studies.

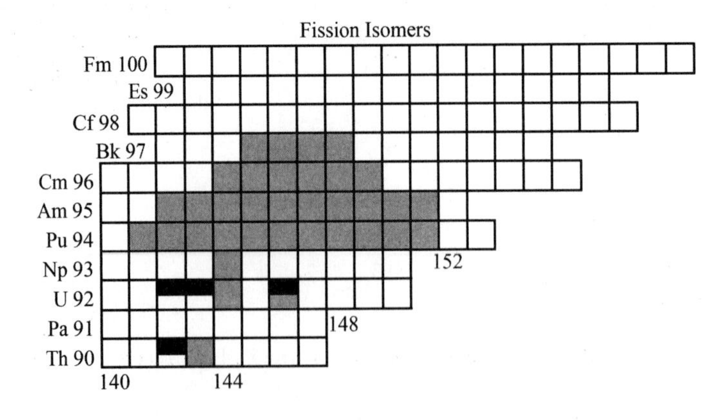


**Figure 4.2 Position of the known spontaneously fissioning isomers in the nuclide chart**

## 4.3 Induced Fission

The absorption of a neutron by most isotopes involves radioactive capture, with the excitation energy appearing as a gamma ray. In certain heavy elements, notably uranium and plutonium, an alternate consequence is observed—the splitting of the nucleus into two massive fragments, a process called fission. Figure 4.3 shows the sequence of events, using the reaction with $^{235}U$ to illustrate. In Stage A, the neutron approaches the $^{235}U$ nucleus. In Stage B, the $^{236}U$ nucleus has been formed, in an excited state. The excess energy in some cases may be released as a gamma ray, but more frequently, the energy causes distortions of the nucleus into a dumbbell shape, as in Stage C. The parts of the nucleus oscillate in a manner analogous to the motion of a drop of liquid. Because of the dominance of electrostatic repulsion over nuclear attraction, the two parts can separate, as in Stage D. They are then called fission fragments, bearing most of the energy released.

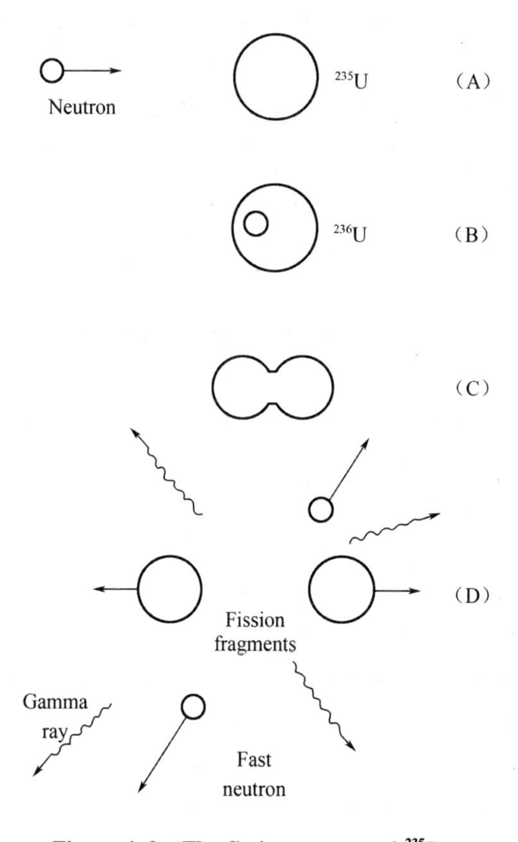


**Figure 4.3 The fission process of $^{235}U$**

They fly apart at high speeds, carrying some 166 MeV of kinetic energy out of the total of around 200 MeV released in the whole process. As the fragments separate, they lose atomic electrons, and the resulting high-speed ions lose energy by interaction with the atoms and molecules of the surrounding medium. The resultant thermal energy is recoverable if the fission takes place in a nuclear reactor. Also shown in the diagram are the prompt gamma rays and fast neutrons that are released at the time of splitting.

The nuclear reaction equation for fission resulting from neutron absorption in $^{235}U$ may be written in general form, letting the chemical symbols for the two fragments be labelled F1 and F2 to indicate many possible ways of splitting. Thus

$$^{235}_{92}U + ^{1}_{0}n \longrightarrow ^{A1}_{Z1}F_1 + ^{A2}_{Z2}F_2 + \upsilon ^{1}_{0}n + E.$$

The appropriate mass numbers and atomic numbers are attached. One example, in which the fission fragments are isotopes of krypton and barium, is

$$^{235}_{92}U + ^{1}_{0}n \longrightarrow ^{90}_{36}Kr + ^{144}_{56}Ba + 2^{1}_{0}n + E.$$

Mass numbers ranging from 75 to 160 are observed, with the most probable at around 92 and 144.

As a collection of isotopes, these byproducts are called fission products. The isotopes have an excess of neutrons or a deficiency of protons in comparison with naturally occurring elements. For example, the main isotope of barium is $^{137}_{56}Ba$, and a prominent element of mass 144 is $^{144}_{60}Nd$. Thus there are seven extra neutrons or four too few protons in the barium isotope from fission, and it is highly unstable. Radioactive decay, usually involving several emissions of beta particles and delayed gamma rays in a chain of events, brings the particles down to stable forms. An example is

$$^{90}_{36}Kr \xrightarrow[33\ s]{} ^{90}_{37}Rb \xrightarrow[2.91\ min]{} ^{90}_{38}Sr \xrightarrow[27.7\ a]{} ^{90}_{39}Y \xrightarrow[64\ h]{} ^{90}_{40}Zr.$$

The hazard associated with the radioactive emanations from fission products is evident when we consider the large yields and the short half-lives.

## 4.4 Energy Considerations

The absorption of a neutron by a nucleus such as $^{235}U$ gives rise to extra internal energy of the product, because the sum of masses of the two interacting particles is greater than that of a normal $^{236}U$ nucleus. We write the first step in the reaction

$$^{235}_{92}U + ^{1}_{0}n \longrightarrow (^{236}_{92}U)^{*},$$

where the asterisk signifies the excited state. The mass in atomic mass units of $(^{236}U)^{*}$ is the sum 235.043 925+1.008 665 = 236.052 590. However, $^{236}U$ in its ground state has a mass of only 236.045 563, lower by 0.007 027 u or 6.5 MeV. This amount of excess energy is sufficient to cause fission. Figure 4.4 shows these energy relationships.

The above calculation did not include any kinetic energy brought to the reaction by the neutron, on the ground that fission can be induced by absorption in $^{235}U$ of very slow neutrons. Only one natural isotope, $^{235}_{92}U$, undergoes fission in this way, while $^{239}_{94}Pu$ and $^{233}_{92}U$ are the main artificial isotopes that do so. Most other heavy isotopes require significantly larger excitation energy to bring the compound nucleus to the required energy level for fission to occur, and the extra energy must be provided by the motion of the incoming neutron. For example, neutrons of at least 0.9 MeV are required to cause fission from $^{238}U$, and other isotopes require even higher energy. The precise terminology is as follows: are those giving rise to fission with slow neutrons;

many isotopes are fissionable if enough energy is supplied. It is advantageous to use fast neutrons—of the order of 1 MeV energy—to cause fission. As will be discussed later, the fast reactor permits the "breeding" of nuclear fuel. In a few elements such as californium, spontaneous fission take place. The isotope, produced artificially by a sequence of neutron absorption, has a half-life of 2.646 a, decaying by alpha emission (97%) and spontaneous fission (3%).

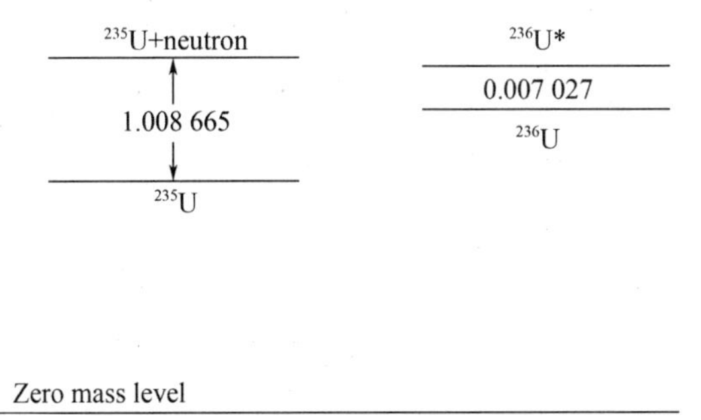


**Figure 4.4 Excitation energy due to neutron absorption**

It may be surprising that the introduction of only 6.5 MeV of excitation energy can produce a reaction yielding as much as 200 MeV. The explanation is that the excitation triggers the separation of the two fragments and the powerful electrostatic force provides them a large amount of kinetic energy. By conservation of mass-energy, the mass of the nuclear products is smaller than the mass of the compound nucleus from which they emerge.

## 4.5 Fission Product Mass Distributions

One of the first big surprises in early studies of fission was the fission product mass distribution. Investigations of the thermal neutron-induced fission of uranium and plutonium nuclides (and later the spontaneous fission of $^{252}$Cf) showed the most probable division of mass was asymmetric ($M_H/M_L$ = 1.3−1.5). The liquid drop model would predict that the greatest energy release and, therefore, the most probable mass split, would be a symmetric one, that is, $M_H/M_L$ = 1.0. This situation is shown in Figure 4.5 where the mass distributions for the thermal neutron induced fission of the "big three nuclides", $^{233}$U, $^{235}$U, and $^{239}$Pu, are shown. Symmetric fission is suppressed by at least two orders of magnitude relative to asymmetric fission.

The key to understanding this situation can be seen in Figures 4.5 and 4.6. In these figures, it is shown that as the mass of the fissioning system increases, the position of the heavy peak in the fission mass distribution remains constant while the position of the light peak increases with increasing fissioning system mass. This observation, along with the observation that the lower edge of the heavy fragment peak is anchored at $A$ = 132 has suggested that the preference for asymmetric fission is due to the special stability of having one fragment with $Z$ = 50, $N$ = 82, a doubly magic spherical nucleus.

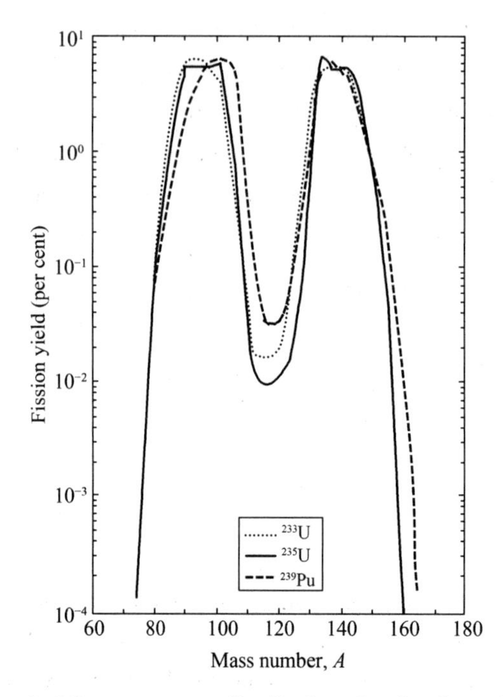


**Figure 4.5 Smoothed fragment mass distributions for the thermal neutron-induced fission of $^{233}U$, $^{235}U$, and $^{239}Pu$. [From Seaborg and Loveland (1990).]**

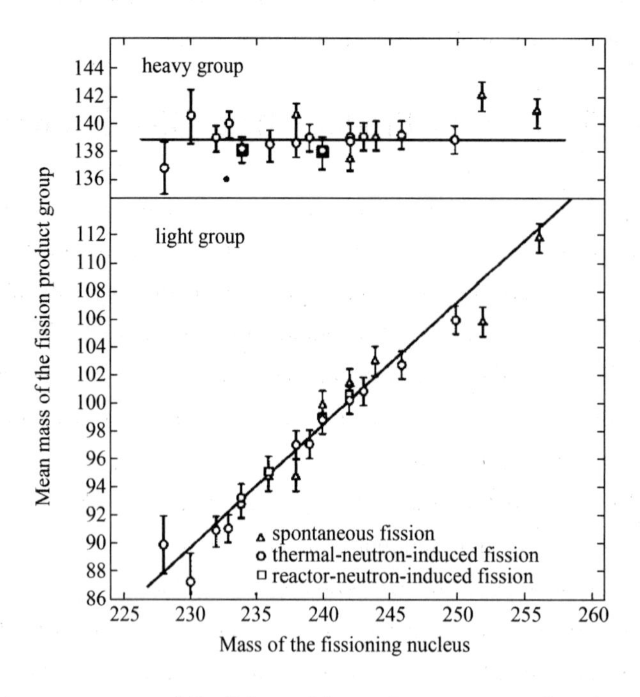


**Figure 4.6 Average masses of the light and heavy fragments as a function of the mass of the fissioning system. [From K. Flynn, et al., Phys. Rev. C, 5, 1728 (1972). Copyright (1972) by the American Physical Society.]**

## 词汇

| | |
|---|---|
| fission process | 裂变过程 |
| schematic view | 示意图 |
| deformation | 形变 |
| transition state | 过渡态 |
| saddle point | 鞍点 |
| surface energy | 表面能 |
| rate-determining step | 决速步 |
| activation energy barrier | 活化能垒 |
| fragment | 碎片 |
| scission point | 断点 |
| repulsion | 斥力 |
| accelerate | 加速 |
| emission | 发射 |
| prompt neutron | 瞬发中子 |
| delayed neutron | 缓发中子 |
| spontaneous fission | 自发裂变 |
| atomic number | 原子序数 |
| half-life | 半衰期 |
| short-lived | 短寿命的 |
| barrier penetration | 势垒穿透 |
| fission barrier | 裂变势垒 |
| frequency | 频率 |
| thorium | 钍 |
| berkelium | 锫 |
| moments of incrtia | 惯性矩 |
| rotational band | 转动谱带 |
| quadrupole moment | 四极矩 |
| induced fission | 诱发裂变 |
| distortion | 畸变 |
| oscillate | 振荡 |
| recoverable | 可回收的 |
| split | 分裂 |
| krypton | 氪 |

| | |
|---|---|
| barium | 钡 |
| byproduct | 副产品 |
| large yield | 大的产额 |
| asterisk | 星号 |
| artificial isotope | 人造核素 |
| terminology | 术语 |
| breeding | 增殖 |
| californium | 锎 |
| mass distribution | 质量分布 |
| asymmetric | 不对称的 |
| symmetric | 对称的 |
| doubly magic spherical nucleus | 双幻核 |

## 注释

1. A nucleus with some equilibrium deformation absorbs energy, becoming excited, and deforms to a configuration known as the transition state or saddle point configuration.

具有某种平衡形变的原子核吸收能量被激发,并变形成一种过渡态构型,或称为鞍点构型。

2. As it deforms, the nuclear Coulomb energy decreases (as the average distance between the nuclear protons increases) while the nuclear surface energy increases (as the nuclear surface area increases).

随着它的变形,核库仑能减少(随着核质子之间的平均距离增加),而核表面能增加(随着核表面积的增加)。

3. As these accelerated primary fragments move away from one another, they contract to more spherical shapes, converting their potential energy of deformation into internal excitation energy, that is, they get "hot".

当这些加速的原生碎片相互远离时,它们会收缩成更多的球形,将其变形的势能转化为内部激发的能量,也就是说,它们会变得"热"。

4. This excitation energy is removed by the emission of the "prompt" neutrons from the fully accelerated fragments and then, in competition with the last neutrons to be emitted, the nucleus emits gamma rays.

这一激发能量被来自完全加速的碎片的"瞬发"中子的发射所消除,然后在与最后发射的中子的竞争中,原子核发射出 γ 射线。

5. γ-ray decay back to the ground state limits the number of isomers with lower $Z$ and $N$ than those in this island, whereas spontaneous fission decay limits the number of cases with high $Z$ and $N$.

γ 射线衰变回基态限制了具有较低 $Z$ 和 $N$ 的同分异构体的数量,而自发裂变衰变限制了具有较高 $Z$ 和 $N$ 的同分异构体的数量。

6. Spectroscopic studies of the transitions between the states in the second minimum have shown that the moments of inertia associated with the rotational bands are those expected for an object with an axes ratio of 2 : 1—a result confirmed in quadrupole moment studies.

对第二种最小状态之间的过渡的光谱研究表明,与旋转谱带相关的惯性矩是轴比为 2 : 1 的物体所预期的惯性矩,这一结果在四极矩研究中得到了证实。

7. The excess energy in some cases may be released as a gamma ray, but more frequently, the energy causes distortions of the nucleus into a dumbbell shape, as in Stage C.

在某些情况下,多余的能量可能以 γ 射线的形式释放出来,但更常见的是,能量使原子核扭曲成哑铃形状,如阶段 C。

8. The nuclear reaction equation for fission resulting from neutron absorption in $^{235}U$ may be written in general form, letting the chemical symbols for the two fragments be labelled F1 and F2 to indicate many possible ways of splitting.

在 $^{235}U$ 中吸收中子而产生裂变的核反应方程式可以用一般的形式来写,让两个碎片的化学符号标为 F1 和 F2,以表示许多可能的分裂方式。

9. Radioactive decay, usually involving several emissions of beta particles and delayed gamma rays in a chain of events, brings the particles down to stable forms.

在一系列(衰变)事件中,放射性衰变通常通过多次释放 β 粒子和延迟的 γ 射线,将粒子降低到稳定的形式。

10. The hazard associated with the radioactive emanations from fission products is evident when we consider the large yields and the short half-lives.

当我们考虑到大的产额和短的半衰期时,与裂变产物的放射性辐射有关的危害是显而易见的。

11. The absorption of a neutron by a nucleus such as $^{235}U$ gives rise to extra internal energy of the product, because the sum of masses of the two interacting particles is greater than that of a normal $^{236}U$ nucleus.

像$^{235}U$ 这样的原子核对中子的吸收会产生额外的内能,因为两个相互作用的粒子的质量总和大于正常的$^{236}U$ 原子核的质量总和。

12. Most other heavy isotopes require significantly larger excitation energy to bring the compound nucleus to the required energy level for fission to occur, and the extra energy must be provided by the motion of the incoming neutron.

大多数其他重质同位素需要明显更大的激发能量,以使复合核达到发生裂变所需的能级,而额外的能量必须由进入的中子的运动提供。

13. The explanation is that the excitation triggers the separation of the two fragments and the powerful electrostatic force provides them a large amount of kinetic energy.

因为激发引发了两个碎片的分离,强大的静电力为它们提供了大量的动能。

14. Investigations of the thermal neutron-induced fission of uranium and plutonium nuclides (and later the spontaneous fission of $^{252}Cf$) showed the most probable division of mass was asymmetric.

对铀和钚核素的热中子诱导裂变(以及后来$^{252}Cf$ 的自发裂变)的调查表明,最可能的质量划分是不对称的。

15. In these figures, it is shown that as the mass of the fissioning system increases, the position of the heavy peak in the fission mass distribution remains constant while the position of the light peak increases with increasing fissioning system mass.

这些数字表明,随着裂变系统质量的增加,裂变质量分布中重峰的位置保持不变,而轻峰的位置随着裂变系统质量的增加而增加。

16. This observation, along with the observation that the lower edge of the heavy fragment peak is anchored at $A=132$ has suggested that the preference for asymmetric fission is due to the special stability of having one fragment with $Z=50$, $N=82$, a doubly magic spherical nucleus.

这一观察结果,以及重碎片峰下缘位于 $A=132$ 的观察结果表明,不对称裂变的偏好是由于 $Z=50$、$N=82$ 的一个双幻核碎片具有特殊的稳定性。

# Chapter 5 Natural Radioelements and Natural Decay Chains

## 5.1 Natural Radioelements

Natural radioelements are those elements present in nature that have no stable isotopes, as listed in Table 5.1. They contribute to the environmental radioactivity and their distribution within the earth has many implications in geology and geochemistry. These radioelements include polonium, astatine, radon, francium, actinium, radium, thorium, protactinium, and uranium and have atomic numbers ranging from 84 to 92. Astatine is added since it is formed in rare branchings of natural Po isotopes, but this radioelement is generally classed with the synthetic ones; its total amount in earth's crust is less than 30 g. The list deliberately does not include technetium and promethium, although these are produced in minute amounts in the spontaneous fission of uranium, nor neptunium and plutonium found in uranium ores as the result of natural nuclear reactions; in the present context they are considered as synthetic. Heavy natural radionuclides include isotopes of the stable elements thallium, lead, and bismuth.

**Table 5.1 The natural radioelements**

| Atomic number $Z$ | Name of the element (symbol) | Longest-lived nuclide (half-life) | Discovery | Remarks |
|---|---|---|---|---|
| 84 | Polonium (Po) | $^{209}$Po(102 a) | 1898, P. and M. Curie | Similar to Te |
| 85 | Astatine (At) | $^{210}$At (8.3 h) | 1940, Corson, McKenzie, and Segrè | Halogen; volatile |
| 86 | Radon (Rn) | $^{222}$Rn (3.825 d) | 1900, Rutherford and Soddy | Noble gas |
| 87 | Francium (Fr) | $^{223}$Fr (21.8 min) | 1939, Perey | Alkali metal; similar to Cs |
| 88 | Radium (Ra) | $^{226}$Ra (1 600 a) | 1898, P. and M. Curie | Alkaline earth metal; similar to Ba |
| 89 | Actinium (Ac) | $^{227}$Ac (21.773 a) | 1899, Debierne | Similar to La; more basic |
| 90 | Thorium (Th) | $^{232}$Th ($1.405\times10^{10}$ a) | 1828, Berzelius | Only in the oxidation state Ⅳ; similar to Ce(Ⅳ), Zr(Ⅳ), and Hf(Ⅳ); strongly hydrolyzing; many complexes |

**Table 5.1**(Continued)

| Atomic number $Z$ | Name of the element (symbol) | Longest-lived nuclide (half-life) | Discovery | Remarks |
|---|---|---|---|---|
| 91 | Protactinium(Pa) | $^{231}$Pa ($3.276\times10^4$ a) | 1917, Hahn and Meitner | Preferably in the oxidation state V; very strongly hydrolyzing; many complexes |
| 92 | Uranium (U) | $^{238}$U ($4.468\times10^9$ a) | 1789, Klaproth | Preferably in the oxidaton states Ⅳ and Ⅵ; in solution $UO_2^{2+}$ ions; many complexes |

These elements would not exist on earth without the occurrence of an island of stability formed by the three radionuclides $^{238}$U ($T = 4.468\times10^9$ a), $^{235}$U ($T = 0.704\times10^9$ a), and $^{232}$Th ($T = 14.05\times10^9$ a). These half-lives are sufficiently long to allow survival since the formation of the earth, $4.5\times10^9$ a ago, and correspond to the definition of primordial radionuclides.

The lowest limit of half-life which would prevent the complete extinction of a primordial radionuclide is roughly estimated in the following way. The mass of the earth is $6\times10^{27}$ g. If it were assumed that the earth at the time of its formation was composed of only one type of radionuclide of mass 250, the initial number of atoms would have been $1.5\times10^{49}$. If the half-life were $0.027\times10^9$ a, only about one or several atoms would survive today. This value may be compared to the half-lives of the longest-lived known heavy elements other than the three previous mentioned, i.e., $^{247}$Cm ($T=0.016\times10^9$ a), $^{236}$U ($T=0.02\times10^9$ a), and $^{244}$Pu ($T=0.083\times10^9$ a). Only the latter might have had a slight chance of survival.

This nuclide was first mentioned as an extinct one in order to explain the excess of $^{134}$Xe and $^{136}$Xe in some meteorites. To get a more realistic figure for the amount of natural $^{244}$Pu which could be found on earth, it is assumed that, at the formation of the solar system (at a time very close to the age of the earth), the ratio $^{244}$Pu/$^{238}$U was 0.5 and that a total amount of $4.08\times10^{13}$ t of U was located in the earth's crust. This gives a ratio $^{244}$Pu/$^{238}$U $= 0.9\times10^{-16}$, which leads to 4 kg of surviving $^{244}$Pu. If this Pu is homogeneously distributed in the upper layer of the crust, the concentration would be about 10 atoms per cubic centimeter. The reputed discovery of $10^{-18}$ g of $^{244}$Pu/g in the mineral bastnaesite would point to a $10^6$ to $10^8$-fold enrichment.

It will be noted that the half-life of $^{235}$U does not substantially exceed the limit of $0.03\times10^9$ a. The present isotopic abundance of $^{235}$U is about 0.71%; at the time of formation of the earth it was 28% and $2\times10^9$ a ago it was about 10%, a proportion high enough to trigger a fission chain reaction in a natural nuclear reactor.

## 5.2 Natural Radioactive Families

The natural radioactive chains are represented in Figures 5.1–5.4. The dominant steps in the chains are α and β decays, which either decrease the mass number by 4 or leave it unchanged. Therefore, there are four chains (including one which is extinct on Earth) corresponding to mass numbers $A = 4n$ ($^{232}$Th and descendants) through $A = 4n + 3$ ($^{235}$U and descendants), where $n$ is an integer.

### 4*n* Chain

This chain is defined in Figure 5. 1. Because α decay changes mass number by 4, the overall loss of 24 mass numbers between $^{232}$Th and $^{208}$Pb requires six α's. But six α's would change the atomic number by 12, and the overall change is only 8; there must be a net of four β decays (β decays less electron captures), each increasing atomic number by 1 without changing mass number. In fact, there are no electron capture decays in any of the chains—for reasons related to general trends in nuclear stability—and there are exactly four β decays in this chain.

| | | | | | | |
|---|---|---|---|---|---|---|
| | | | | | $^{236}$U<br>2.3×10$^{7}$<br>years | |
| | | | | | 100<br>4494<br>4445 | |
| | | RdTh | $^{228}$Th<br>1.91 years | | $^{232}$Hh<br>1.4×10$^{10}$<br>years | |
| | | $MsTh_2$ | 100<br>5423<br>5340 | $^{228}$Ac<br>6.1 h | 100<br>4013<br>3950 | |
| | | ThX | $^{224}$Ra<br>3.64 days | | $^{228}$Ra<br>5.76 years | $MsTh_1$ |
| | | | 100<br>5686<br>5449 | | | |
| | | Tn | $^{220}$Rn<br>56 s | | | |
| | | | 100<br>6288 | | | |
| ThC′ | $^{212}$Po<br>0.3 μs | | $^{216}$Po<br>0.15 s | ThA | | |
| ThC | 100<br>8784 | $^{212}$Bi<br>61 min | 100<br>6779 | | | β |
| | $^{208}$Pb | 35.9<br>6051 | $^{212}$Pb<br>10.6 h | ThB | | α |
| ThC″ | | $^{208}$Tl<br>3.1 min | | | | |

**Figure 5. 1 4*n* Chain**

The parent of the 4$n$ series, 1. 405×10$^{10}$ a $^{232}$Th, is the longest-lived nuclide in the Th-U region. It is comparable to the age of the universe, so more than half of all of the $^{232}$Th ever produced (and not consumed by nuclear reactions) is still present. Because of the long half-life,

$^{232}Th$ has unusually low specific activity. Little easily absorbed radiation, such as α particles and low-energy electrons and photons, escapes the necessarily thick samples of moderate activity.

The longest-lived member in the family is $^{228}Ra$ (5.76 a), but at equilibrium less than 1 mg exists in 1 t of thorium (Table 5.2). The last steps in the chain are unusual; $^{212}Po$ has a remarkably short half-life (and high decay energy), and $^{208}Tl$ emits a high energy gamma ray (2 615 keV). This gamma ray is a distinctive feature of background spectra of most gamma ray spectro-meters.

**Table 5.2 Amounts of nuclides in equilibrium with 1 t of natural $^{232}Th$ and estimated amounts in the earth's crust**

| Nuclide | Half-life | Amount in 1 t U | Amount in earth's crust |
|---|---|---|---|
| $^{228}Ra$ | 5.76 a | 40 2 μg | $10^5$ t |
| $^{228}Ac$ | 6.13 h | 48.6 ng | 12 t |
| $^{228}Th$ | 1.913 a | 133 μg | $3.3\times10^4$ t |
| $^{224}Ra$ | 3.66 d | 690 ng | 172 t |
| $^{220}Rn$ | 55 s | 117 pg | 29 kg |
| $^{216}Po$ | 0.15 s | 620 fg | 155 g |
| $^{212}Pb$ | 10.64 h | 79 ng | 20 t |
| $^{212}Bi$ | 60.6 min | 7.5 ng | 1.9 t |
| $^{212}Po$ | $3\times10^{-7}$ s | $4\times10^{-19}$ g | 100 μg |
| $^{208}Tl$ | 3.05 min | 133 pg | 33 g |

### 4*n*+2 Chain

This Chain is defined in Figure 5.2. The $^{238}U$ family ends with the stable $^{206}Pb$ after 8 α and 6 $\beta^-$ disintegrations. The longest-lived members are $^{234}U$ ($T=0.245\times10^6$ a), which is the third natural uranium isotope (abundance 0.005 5% or 55ppm), and $^{230}Th$ ($T=0.08\times10^6$ a). The weights of nuclides contained at present in 1 t of natural uranium (992.8 kg of $^{238}U$) are given in Table 5.3. They are inversely proportional to the half-lives. Apart from U and Th, only Ra and $^{210}Pb$ are present in milligram amounts. In this family are found the longest-lived known isotopes of the elements U, Ra, Rn, and the longest-lived natural isotope of Po.

The values in Table 5.3 are calculated for the state of radioactive equilibrium. In rocks and minerals, equilibria are frequently perturbed by recoil effects and selective lixiviation. The precursors of Ra are soluble in groundwaters with a high content in $SO_4^{2-}$ ions and are insoluble in media rich in $Cl^-$, whereas Ra and its daughters have the opposite behavior. Even the activity ratio of the uranium isotopes $^{234}U$ and $^{238}U$ may differ markedly from the theoretical value of unity in minerals and waters. The total activity per unit mass of a uraniferous deposit, with a given U content, can vary greatly and members of the series can migrate to long distance from the parent.

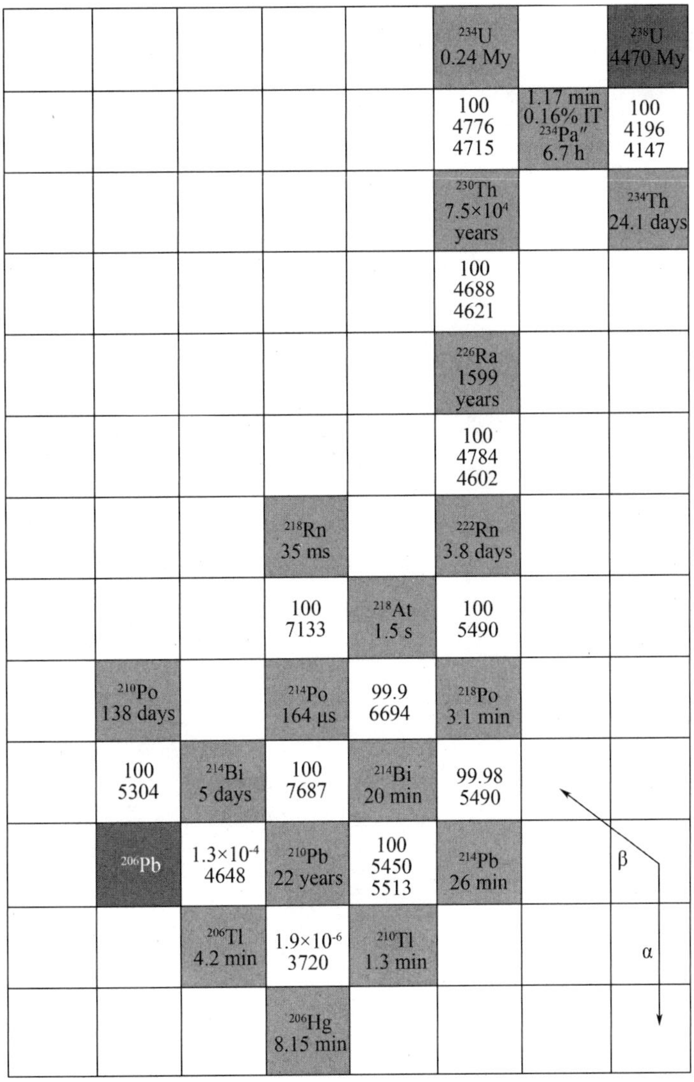

| | | | | | | | |
|---|---|---|---|---|---|---|---|
| | | | | | $^{234}U$<br>0.24 My | | $^{238}U$<br>4470 My |
| | | | | | 100<br>4776<br>4715 | 1.17 min<br>0.16% IT<br>$^{234}Pa''$<br>6.7 h | 100<br>4196<br>4147 |
| | | | | | $^{230}Th$<br>$7.5\times10^4$<br>years | | $^{234}Th$<br>24.1 days |
| | | | | | 100<br>4688<br>4621 | | |
| | | | | | $^{226}Ra$<br>1599<br>years | | |
| | | | | | 100<br>4784<br>4602 | | |
| | | | $^{218}Rn$<br>35 ms | | $^{222}Rn$<br>3.8 days | | |
| | | | 100<br>7133 | $^{218}At$<br>1.5 s | 100<br>5490 | | |
| | $^{210}Po$<br>138 days | | $^{214}Po$<br>164 μs | 99.9<br>6694 | $^{218}Po$<br>3.1 min | | |
| | 100<br>5304 | $^{214}Bi$<br>5 days | 100<br>7687 | $^{214}Bi$<br>20 min | 99.98<br>5490 | | |
| | $^{206}Pb$ | $1.3\times10^{-4}$<br>4648 | $^{210}Pb$<br>22 years | 100<br>5450<br>5513 | $^{214}Pb$<br>26 min | | β |
| | | $^{206}Tl$<br>4.2 min | $1.9\times10^{-6}$<br>3720 | $^{210}Tl$<br>1.3 min | | | α |
| | | | $^{206}Hg$<br>8.15 min | | | | |

**Figure 5.2 4*n*+2 Chain**

**Table 5.3 Amounts of nuclides in equilibrium with $^{238}U$ in 1 ton of natural uranium and estimated amounts in the earth's crust**

| Nuclide | Half-life | Amount in 1 ton U | Amount in earth's crust |
|---|---|---|---|
| $^{234}Th$ | 24.1 d | 14 μg | $1.4\times10^3$ t |
| $^{234m}Pa$ | 1.17 min | 0.42 ng | 42 kg |
| $^{234}U$ | $2.45\times10^5$ a | 53.5 g | $5\times10^9$ t |
| $^{230}Th$ | $8\times10^4$ a | 17.25 g | $2\times10^9$ t |
| $^{226}Ra$ | $1.6\times10^3$ a | 337 mg | $3.4\times10^7$ t |
| $^{222}Rn$ | 3.823 d | 2.17 μg | 210 t |
| $^{218}Po$ | 3.05 min | 1.2 ng | 120 kg |
| $^{214}Pb$ | 26.8 min | 10.1 ng | 1 ton |

**Table 5.3**(Continued)

| Nuclide | Half-life | Amount in 1 ton U | Amount in earth's crust |
|---|---|---|---|
| $^{214}Bi$ | 19.7 min | 7.5 ng | 750 kg |
| $^{214}Po$ | $1.64\times10^{-4}$ s | 1 fg | 100 mg |
| $^{210}Pb$ | 22.3 a | 4.37 mg | $4.3\times10^5$ t |
| $^{210}Bi$ | 5 d | 2.7 μg | 270 t |
| $^{210}Po$ | 138.4 d | 74 μg | $7.4\times10^3$ t |

### 4*n*+3 Chain

This Chain is defined in Figure 5.3. The $4n+3$ or actinium series has $^{235}U$ as the parent and $^{207}Pb$ as the stable end product resulting from 7 α and 4 $\beta^-$ decays. This family is peculiar in that it includes the three longest-lived isotopes of Pa, Ac, and Fr, and the longest-lived natural radioactive Tl isotope ($^{207}Tl$, $T_{1/2}=4.77$ min). Fr exists solely in this family. The only member present in ponderable amount in 1 t of U is $^{231}Pa$ (see Table 5.4).

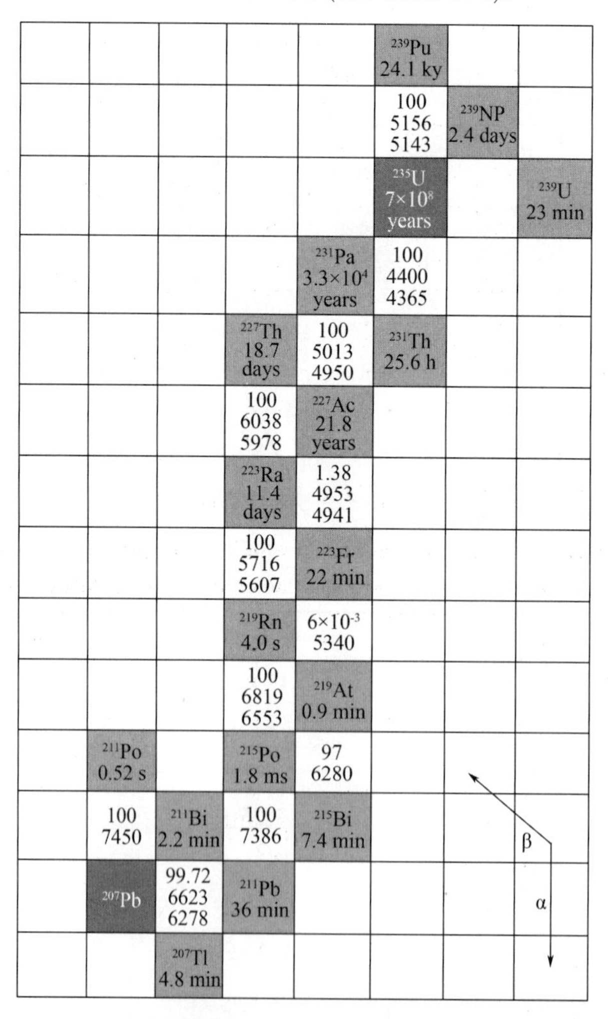


**Figure 5.3  4*n*+3 Chain**

**Table 5.4 Amounts of nuclides in equilibrium with $^{235}U$ in 1 t of natural uranium and estimated amounts in the earth's crust**

| Nuclide | Half-life | Amount in 1 t U | Amount in earth's crust |
|---|---|---|---|
| $^{231}Th$ | 1.06 d | 29.9 ng | 3 t |
| $^{231}Pa$ | $3.25 \times 10^4$ a | 327 mg | $3.2 \times 10^7$ t |
| $^{227}Ac$ | 21.7 a | 211 μg | $2.1 \times 10^4$ t |
| $^{227}Th$ | 18.7 d | 506 ng | 50 t |
| $^{223}Fr$ | 22 min | 5.5 pg | 550 g |
| $^{223}Ra$ | 11.4 d | 303 ng | 30 t |
| $^{219}Rn$ | 3.96 s | 1.2 pg | 120 g |
| $^{215}Po$ | $1.8 \times 10^{-3}$ s | 0.5 fg | 50 mg |
| $^{211}Pb$ | 36.1 min | 630 pg | 63 kg |
| $^{211}Bi$ | 2.14 min | 37.4 pg | 3.7 kg |
| $^{207}Tl$ | 4.77 min | 79 pg | 7.9 kg |

U is widely dispersed in the environment, where it is found in the Earth's crust in the form of U(Ⅳ) and U(Ⅵ) at an average concentration of 3ppm (estimated limits are 1ppm to 10ppm in igneous rocks), and in the oceans in the oxidized form U(Ⅵ) to the extent of $10^{-3}$ppm (limits $10^{-5}$ppm to $10^{-2}$ppm, depending on the salinity). The U content of the earth's crust to a depth of 20 km is estimated at $10^{14}$ t and the content of the oceans at 1 010 t. Today, only ores containing at least 300 g/t are of economic value; locally, the amount of U may reach concentrations up to 700 kg/t.

### 4*n*+1 Chain

This Chain is defined in Figure 5.4. The $4n+1$ series is extinct on Earth except for vanishing small amounts arising from capture of stray neutrons by $^{232}Th$. Even the end-product (about $2 \times 10^{19}$ a $^{209}Bi$) appears to be slightly radioactive. The longest-lived member is $^{237}Np$, but the most important is $^{233}U$. This isotope of uranium has fission properties similar to $^{235}U$ (and $^{239}Pu$). The mechanisms for large-scale production are similar to those for $^{239}Pu$.

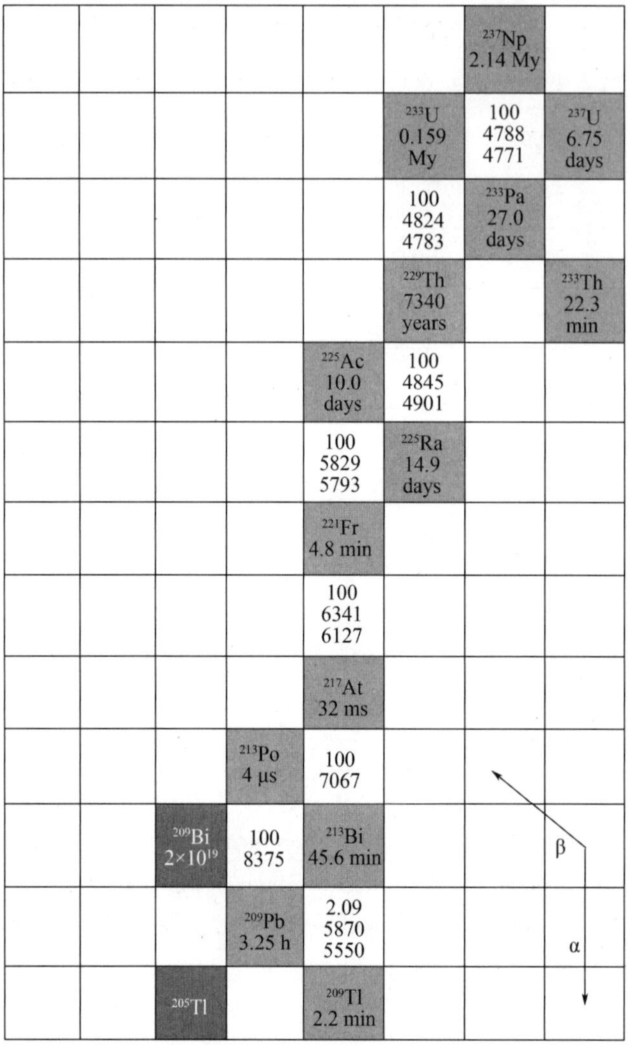

| | | | | | | $^{237}Np$<br>2.14 My | |
|---|---|---|---|---|---|---|---|
| | | | | | $^{233}U$<br>0.159<br>My | 100<br>4788<br>4771 | $^{237}U$<br>6.75<br>days |
| | | | | | 100<br>4824<br>4783 | $^{233}Pa$<br>27.0<br>days | |
| | | | | | $^{229}Th$<br>7340<br>years | | $^{233}Th$<br>22.3<br>min |
| | | | | $^{225}Ac$<br>10.0<br>days | 100<br>4845<br>4901 | | |
| | | | | 100<br>5829<br>5793 | $^{225}Ra$<br>14.9<br>days | | |
| | | | | $^{221}Fr$<br>4.8 min | | | |
| | | | | 100<br>6341<br>6127 | | | |
| | | | | $^{217}At$<br>32 ms | | | |
| | | | $^{213}Po$<br>4 μs | 100<br>7067 | | | |
| | | $^{209}Bi$<br>$2\times10^{19}$ | 100<br>8375 | $^{213}Bi$<br>45.6 min | | | β |
| | | | $^{209}Pb$<br>3.25 h | 2.09<br>5870<br>5550 | | | α |
| | | $^{205}Tl$ | | $^{209}Tl$<br>2.2 min | | | |

**Figure 5.4 4*n*+1 Chain**

# 5.3 Atmospheric Radioactivity

Natural radioactivity in the atmosphere is due mainly to the presence of one Rn isotope in each family: $^{222}Rn$ ($T$=3.8 d), $^{220}Rn$ ($T_{1/2}$=55 s) and $^{219}Rn$ ($T$=4 s). Only the first two have lifetimes that are compatible with their release from soils and waters. The most important is $^{222}Rn$ because its relatively long half-life ensures a higher steady concentration in the environment. The rate of release from the surface of the ground is about one atom $cm^{-2} \cdot s^{-1}$.

The concentration of Rn in air varies within several orders of magnitude, depending on the location. The total activity of the atmosphere is $10^{18}$ Bq (0.8 mol) and its average concentration near the surface is several Bq $m^{-3}$, corresponding to a few atoms per cubic centimeter. Values between 15 and 70 Bq $m^{-3}$ have been reported for air in dwellings. In certain caves the radon concentration may reach 7 500 Bq $m^{-3}$.

The descendants of radon isotopes are solids, which either condense on aerosols that are always present in the atmosphere, or react with atmospheric gases. Eventually, they fall to the surface of the earth and are thus also taken up by ocean waters. The behavior of radon isotopes is used for the study of atmospheric circulations.

## 5.4 Radionuclides Found in Nature

### Primary radionuclides

Long-lived radionuclides formed in the process of nucleogenesis are called primary or primordial radionuclides. The most important is $^{40}K$, half-life $1.25 \times 10^9$ a. Its present isotopic abundance is 0.011 7% as compared to 0.145% in the early earth. Potassium is highly abundant in nature.

About 20 primary radionuclides have been recorded. They possess very long $\beta^-$ half-lives and $\alpha$ half-lives. Most of them are very difficult to detect because (1) the isotopic abundance is sometimes below 1 %, (2) the half-lives are associated with low energies of the emitted $\alpha$ and $\beta$ particles, and (3) the $\gamma$ photons are highly converted. For these reasons, the primary radionuclides are of little concern in radiochemistry. They are even considered as stable nuclides in the isotopic composition of the corresponding elements.

### Radionuclides from reactions of cosmic radiation

A dozen radionuclides are produced in the interaction of cosmic radiations with $^{14}N$, $^{16}O$, and $^{40}Ar$ atoms in the upper levels of the atmosphere. The stationary amount is the highest for the long-lived cosmogenic nuclides $^{10}Be$, $^{14}C$, $^{26}Al$, and $^{36}Cl$, but the average volumic activity is very low because of the enormous dilution factor.

The nuclides are formed with an excess of kinetic energy and react like hot atoms, mainly with $O_2$, to form a variety of labelled molecules; some of these are fixed on aerosols. After a variable residence time in the atmosphere, the nuclides reach the surface of the earth in rain and snow. Owing to isotope exchange and biological reactions, $^3H$ and $^{14}C$ are found in many hydrogen and carbon-containing compounds. Activities of the cosmogenic nuclide are very low and difficult to measure under ordinary conditions.

### 词汇

| | |
|---|---|
| natural radioelement | 天然放射性元素 |
| geology | 地质学 |
| geochemistry | 地质化学 |
| polonium | 钋 |

| | |
|---|---|
| astatine | 砹 |
| radon | 氡 |
| francium | 钫 |
| actinium | 锕 |
| radium | 镭 |
| thorium | 钍 |
| protactinium | 镤 |
| uranium | 铀 |
| crust | 地壳 |
| technetium | 锝 |
| promethium | 钷 |
| neptunium | 镎 |
| thallium | 铊 |
| lead | 铅 |
| bismuth | 铋 |
| primordial radionuclide | 原生放射性核素 |
| meteorite | 陨石 |
| mineral | 矿物 |
| bastnaesite | 氟碳铈矿 |
| enrichment | 富集 |
| abundance | 丰度 |
| natural radioactive families | 天然放射性族 |
| integer | 整数 |
| recoil effect | 反冲效应 |
| selective lixiviation | 选择性浸出 |
| precursor | 母体 |
| uraniferous | 含铀的 |
| ore | 矿石 |
| end-product | 最终产品 |
| large-scale | 大规模 |
| dwelling | 住处 |
| descendant | 衍生物 |
| aerosol | 气溶胶 |
| atmospheric circulation | 大气环流 |
| nucleogenesis | 核起源 |

potassium 钾
photon 光子
radiochemistry 放射化学
cosmic radiation 宇宙辐射
cosmogenic nuclide 宇宙成因核素
volumic activity 体积活度
dilution factor 稀释因子

**注释**

1. The list deliberately does not include technetium and promethium, although these are produced in minute amounts in the spontaneous fission of uranium, nor neptunium and plutonium found in uranium ores as the result of natural nuclear reactions; in the present context they are considered as synthetic.

所列出的核素不包括锝和钷,虽然它们是在铀的自发裂变中产生的微量元素,也不包括在铀矿中发现的作为天然核反应产物的镎和钚,但是在目前情况下,它们被认为是合成物。

2. The present isotopic abundance of $^{235}U$ is about 0. 71%; at the time of formation of the earth it was 28% and $2\times10^9$ a ago it was about 10%, a proportion high enough to trigger a fission chain reaction in a natural nuclear reactor.

目前$^{235}U$的同位素丰度约为 0. 71%;在地球形成时,它是 28%。$2\times10^9$ 年前它约为 10%,这一比例高到足以在天然核反应堆中引发裂变链式反应。

3. Because α decay changes mass number by 4, the overall loss of 24 mass numbers between $^{232}Th$ and $^{208}Pb$ requires six α's. But six α's would change atomic number by 12, and the overall change is only 8; there must be a net of four β decays (β decays less electron captures), each increasing atomic number by 1 without changing mass number.

由于 α 衰变使质量数改变 4,因此在 $^{232}Th$ 和 $^{208}Pb$ 之间,质量数总损失为 24,这需要 6 次 α 衰变。6 次 α 衰变会使原子序数改变 12,但是原子序数的变化只有 8,因此必须有 4 次 β 衰变(β 衰变发生电子捕获的情况较少),每次原子序数增加 1 而不改变质量数。

4. The precursors of Ra are soluble in groundwaters with a high content in $SO_4^{2-}$ ions and are insoluble in media rich in $Cl^-$, whereas Ra and its daughters have the opposite behavior.

Ra 的母体可溶于 $SO_4^{2-}$ 离子含量高的地下水,不溶于富含 $Cl^-$的介质,而 Ra 及其子体的行为则相反。

5. The total activity per unit mass of a uraniferous deposit, with a given U content, can vary greatly and members of the series can migrate to long distance from the parent.

在 U 含量一定的情况下,单位质量的含 U 矿床的总活性会有很大变化,U 系核素可以迁移到离母体很远的地方。

6. The total activity of the atmosphere is $10^{18}$ Bq (0. 8 mol) and its average concentration near the surface is several Bq $m^{-3}$, corresponding to a few atoms per cubic centimeter.

大气层的总活性为 $10^{18}$ Bq(0. 8 mol),其在表面附近的平均浓度为几 Bq $m^{-3}$,相当于每立方厘米几个原子。

7. The descendants of radon isotopes are solids, which either condense on aerosols that are always present in the atmosphere, or react with atmospheric gases.

氡同位素的衍生物是固体,它们要么凝结在大气中始终存在的气溶胶上,要么与大气中的气体发生反应。

8. The nuclides are formed with an excess of kinetic energy and react like hot atoms, mainly with $O_2$, to form a variety of labelled molecules; some of these are fixed on aerosols.

这些核素的形成伴随着过剩的动能并且会像热原子一样发生反应,主要是与 $O_2$ 发生反应,形成各种被标记的分子,其中一些被固定在气溶胶上。

# Chapter 6 Artificial Radioelements and Transuranium Elements

## 6.1 Artificial Radioelements

Artificial (also called synthetic or anthropogenic) radioelements are those which have no stable isotopes and do not exist in nature, or at the most only in minute amounts and often as short-lived ephemeral species. They include $_{43}Tc$, $_{61}Pm$, $_{85}At$, and presently all elements from $_{93}Np$ to $_{109}Une$. They are synthesized in nuclear reactions induced by neutrons for elements up to $_{99}Es$ and $_{100}Fm$, and by charged particles for the heavier ones. In many cases these reactions lead to a mixture of isotopes of a given element for which the composition varies widely with the target, the nature and energy of the projectile, and the duration.

All artificial radioelements with $Z<97$ (except At, whose longest-lived isotope is 8.3 h $^{210}At$), i.e., Tc, Pm, Np, Pu, Am, and Cm, are found in large or appreciable amounts in irradiated U or Pu fuels used in nuclear reactors. In a sense, these fuels may be considered as ores of these elements if they were systematically exploited. Like any ore, they contain impurities in addition to the desired elements; in this case, the impurities are the fission products other than Pm and Tc together with vast amounts of uninteresting and unwanted material, such as the uranium isotopes. However, this kind of artificial ore has a very high activity.

Among the artificial radioelements, only Pu is regularly extracted from irradiated fuels, either to serve as nuclear fuel or for military purposes. Other radioelements ($^{237}Np$) and fission products (mainly, $^{137}Cs$ and $^{90}Sr$) are (or have been) occasionally separated. Rather ambitious projects are now being elaborated in countries operating reprocessing plants to recover the actinides from U to Am as well as most of the elements formed in the fission process. Separation of the minor actinides Np, Am, and Cm could markedly improve the management of nuclear wastes.

The stable end products of fission chains include potentially useful and naturally scarce elements such as Pd and Xe and most of the lanthanides. However, some of these elements contain long-lived radioisotopes such as $6.5\times10^6$ a $^{107}Pd$, which may preclude their use in consumer products, although the nuclide mentioned emits only soft $\beta^-$ rays. The isotopic composition and thus the atomic weight of the stable fission elements differ from those of the natural ones, sometimes by more than 1 %. The atomic weight of natural Cs is 132.905; that of fission Cs is 134.92.

The main part of man-made radioelements produced in large amounts is confined in and around nuclear reactors, in facilities for fuel fabrication and in reprocessing plants. Sealed sources of radioelements are widely distributed because of their numerous applications in research, industry, and hospitals, and their use is subject to stringent regulation.

## 6.2 Discovery of Technetium

According to the periodic table, the missing element at atomic number 43 should be between manganese and rhenium with regard to chemical properties. The group that discovered rhenium in 1925 claimed to have detected element 43 in the same minerals. They subsequently were able to isolate weighable quantities of rhenium (more than a gram in 1928), but neither they nor others could confirm detection of the element 43.

Technetium was finally discovered by Perrier and Segre in 1937. Lawrence had bombarded a Mo plate with deuterons in the Berkeley cyclotron for several months prior to shipping the plate to Segre's group in Italy in late December, 1936. Perrier and Segre began their radiochemical studies in late January, more than 6 weeks after the end of bombardment. On the surface exposed to the deuterons, they found "strong activity, chiefly due to very slow electrons" ascribed to "more than one substance of a half-value period of some months" in addition to $^{32}P$.

Subsequent work revealed that most of the Tc activity arose from what are now known to be 61-day $^{95m}Tc$ and 90.1-day $^{97m}Tc$.

Table 6.1 shows a nuclide chart corresponding to charged particle reactions with Mo. Both (d, n) and (d, 2n) reactions are energetically allowed (have energy requirements, $Q$, that are met by the bombarding particle) and lead directly to Tc. For example, the $Q$ values for $^{95}Mo(d, n)^{96}Tc$ and $^{95}Mo(d, 2n)$ $^{95m}Tc$ are $-3.25$ and $-4.74$ MeV, respectively. The $Q$ value for $^{97}Mo$ $(d, 2n)$ $^{97m}Tc$ is $-3.42$ MeV. In addition, (d, p) reactions with $^{98}Mo$ and $^{100}Mo$ produce Mo isotopes that subsequently decay to Tc. Therefore, the bombardments produced many different radioactive products, but after a few months 61-day $^{95m}Tc$ and 90.1-day $^{97m}Tc$ dominated the activity. Note that both of these are isomeric levels. One, $^{95m}Tc$, decays primarily (96%) by electron capture (EC) and subsequent γ-ray emission. The other, $^{97m}Tc$, decays by a highly converted 96.6 keV transition, which presumably produced the "slow electrons" reported.

**Table 6.1 Nuclide chart showing stable molybdenum isotopes and radioactive products from deuteron bombardments**

| $^{92}Tc$ 4.2 min | $^{93}Tc$ 43 min 2.73 h | $^{94}Tc$ 52 min 4.88 h | $^{95}Tc$ 61 d 20 h | $^{96}Tc$ 52 min 4.3 d | $^{97}Tc$ 91 d 4.2 h | $^{98}Tc$ 4.2 My | $^{99}Tc$ 6.01 h 0.21 My | $^{100}Tc$ 15.8 s | $^{101}Tc$ 14.2 min | $^{102}Tc$ 4.4 min 5.3 s |
|---|---|---|---|---|---|---|---|---|---|---|
| $^{91}Mo$ 1.08 min 15.5 min | $^{92}Mo$ 14.84% | $^{93}Mo$ 6.9 h 3.5 ky | $^{94}Mo$ 9.25% | $^{95}Mo$ 15.92% | $^{96}Mo$ 16.68% | $^{97}Mo$ 9.55% | $^{98}Mo$ 24.12% | $^{99}Mo$ 2.75 d | $^{100}Mo$ 9.63% | $^{101}Mo$ 14.6 min |

Stable isotopes are shaded and list natural abundances in percentage. When isomers exist and relative energies of the states are known, the half-life of the isomer is given above that of the ground state.

## 6.3 Discovery of Promethium

In principle, $_{61}$Pm could have been discovered by techniques similar to those used for $_{43}$Tc — bombardment of $_{60}$Nd with charged particles from a cyclotron and demonstration that some of the products were isotopes of element 61. Indeed the bombardments were done, and several radionuclides that might have been isotopes of element 61, for example, "cyclonium" were produced, but the essential chemical proof was lacking. Whereas technetium and its neighbors differ considerably in chemical properties, variations among neighboring rare earths are very slight. This chemical characterization was beyond the capabilities of the groups carrying out the bombardments.

Conventional chemistry is illustrated by the chemistry used to support the claim of "illinium". This report is one of a series of studies in which concentrates rich in one or more rare earths were studied by arc spectroscopy. Harris et al. chose a concentrate prepared by methods that were intermediate between those found to enhance $_{60}$Nd and $_{62}$Sm. The visible and infrared emissions were found to have about 130 lines that were not unique to either Nd or Sm fractions. They found support for their claims in absorption spectra of solutions and L X-ray emission spectra associated with the concentrate.

The chart of the nuclides in the region of promethium is given in Table 6.2. It is obvious now that there is a negligible amount of promethium in natural materials, and if macroscopic amounts were present, the radioactivity would be intense. The isotope with the longest half-life is 17.7 a $^{145}$Pm. If it were present at 1ppm in a 1-g sample, the activity would be 6 MBq.

In order to separate adjacent elements and to isolate very small quantities, a new chemistry was needed. This chemistry, ion-exchange chromatography with synthetic resins, was developed during the Manhattan project in order to study the products of nuclear fission, which include several light rare earths. The technique was borrowed in order to identify element 61.

By mid-1945, over 30 fission products had been characterized as isotopes of yttrium or one of the lanthanides, but few could be definitely assigned by element and mass numbers. The work that ultimately identified element 61 began with a mixture of radioactive materials that concentrated elements between cerium and samarium. This mixture included a soft β emitter with a half-life of about 4 a and a g emitter of about 11 d.

**Table 6.2 Chart of the nuclides in the region of promethium**

| $^{143}$Pm<br>265 d | $^{144}$Pm<br>360 d | $^{145}$Pm<br>17.7 a | $^{146}$Pm<br>5.53 a | $^{147}$Pm<br>2.62 a | $^{148}$Pm<br>41.3 d<br>5.37 d | $^{149}$Pm<br>2.21 d | $^{150}$Pm<br>2.68 h | $^{151}$Pm<br>1.18 d | $^{152}$Pm<br>13.8 min<br>4.1 min | |
|---|---|---|---|---|---|---|---|---|---|---|
| $^{142}$Nd<br>27.2% | $^{143}$Nd<br>12.2% | $^{144}$Nd<br>23.8%<br>$2.1\times10^{15}$ a | $^{145}$Nd<br>8.3% | $^{146}$Nd<br>17.2% | $^{147}$Nd<br>10.98 d | $^{148}$Nd<br>5.7% | $^{149}$Nd<br>1.73 h | $^{150}$Nd<br>5.6% | $^{151}$Nd<br>12.4 min | |

**Table 6.2**(Continued)

| $^{141}Pr$ 100% | | | | | | ↖ 2.25% | | ↖ 1.08% | | ↖ 0.419% |
|---|---|---|---|---|---|---|---|---|---|---|
| $^{140}Ce$ 88.45% | | $^{142}Ce$ 11.11% | | | | | | | | |

The columns contain isotones. The fission yields for $^{235}U$ are given for several mass chains (i. e., isobars situated along the lines shown by the slanted arrows). Stable isotopes are shaded and list natural abundances in percentage.

## 6.4 Discovery of Transuranium Elements

There were at least two difficulties in the discovery of transuranium elements; the first one was the chemical behavior of transuranium elements. Before Seaborg's proposal of the actinide concept, it was believed that thorium, protactinium, uranium, neptunium, plutonium, and the next elements 95 and 96 should be placed as the heavier members of groups 4 through 10. According to the actinide concept, it was expected that the new series should start with actinium as a prototype and end with the filling of the 5f electron shell at element 103, and that the heavier actinides would be trivalent homologs of the lanthanide series in which 4f orbitals were being filled. The second difficulty was the method of identification of transuranium isotopes produced in nuclear reactions. An isotope to be identified had to be isolated before it decayed from large quantities of impurities, especially from fission products.

The first man-made transuranium element, neptunium (Np), was discovered by McMillan and Abelson (1940) while studying the neutron-induced fission of uranium (U). The reaction used in the synthesis was

$$^{238}U(n,\gamma)^{239}U \xrightarrow[23\ min]{\beta^-} {}^{239}Np \xrightarrow[2.3\ d]{\beta^-} \tag{5.1}$$

Neptunium was named after the planet Neptune, the next planet beyond Uranus.

Plutonium (Pu), the second transuranium element, was discovered by Seaborg and coworkers (Seaborg, et al., 1946). The first isotope of Pu was synthesized by using the 60 in. cyclotron for the deutron bombardment of uranium at the University of California, Berkeley,

$$^{238}U(d,2n)^{238}Np \xrightarrow[2.1\ d]{\beta^-} {}^{238}Pu \xrightarrow[87.7\ a]{\alpha} \tag{5.2}$$

The $\alpha$-emitting product was identified as a new element from the study of chemical behavior of this isotope. It was distinctly different from both uranium and neptunium in its redox properties; the 3+ and 4+ valence states were more stable. A second isotope of element 94, $^{239}Pu$, with a half-life of 24 000 a was synthesized immediately as a daughter of $\beta^-$ decay of $^{239}Np$, which confirmed the presence of element 94. The isotope $^{239}Pu$ produced in appreciable amounts in nuclear reactors is of major importance, because of its large fission cross section with thermal neutrons. It was named after the planet Pluto in analogy to uranium and neptunium.

Element 95, americium (Am), and element 96, curium (Cm) were synthesized by Seaborg et al. in 1944 – 1945 (Seaborg, 1945) in bombardments of uranium and plutonium with α particles in a cyclotron:

$$^{238}U(\alpha, n)^{241}Pu \xrightarrow[15\ a]{\beta^-} {}^{241}Am \xrightarrow[433\ a]{\alpha} \tag{5.3}$$

$$^{239}Pu(\alpha, n)^{242}Cm \xrightarrow[163\ d]{\alpha} {}^{238}Pu \tag{5.4}$$

Americium and curium are quite similar to the rare earth elements in their chemical properties; a stable oxidation state is 3+. Thus, it was too difficult for the discovery team to chemically isolate the two elements from each other and from rare earth fission products. Element 95, americium, was named after the Americas in analogy with the naming of its rare earth homolog, europium, and curium was named after Marie and Pierre Curie in analogy with the naming of its homolog, gadolinium, that was named after the Finnish rare earth chemist Johann Gadolin.

The element with the atomic number 97, berkelium (Bk), was produced by irradiation of milligram amounts of $^{241}$Am with α particles at Berkeley:

$$^{241}Am(\alpha, 2n)^{243}Bk \xrightarrow[4.5\ h]{EC,\ \alpha} \tag{5.5}$$

Berkelium was named after the city of Berkeley, California where it was discovered, just as the name terbium derived from Ytterby, Sweden.

Californium was synthesized in 1950 by irradiation of a few micrograms of $^{242}$Cm with α particles:

$$^{242}Cm(\alpha, n)^{245}Cf \xrightarrow[44\ min]{\alpha,\ EC} \tag{5.6}$$

The identification of element 98 was accomplished with a total of only 5 000 atoms. It was named after the state of its discovery, although the chemical analog of element 98 was dysprosium(Dy).

Einsteinium (Es) and fermium (Fm) were identified in 1952 in the radioactive debris from the "Mike" thermonuclear explosion that took place in the Pacific. Ion-exchange separation was applied, and the new elements, einsteinium and fermium, were isolated by processing larger amounts of the radioactive coral material. The elements Es and Fm were named in honor of Albert Einstein and Enrico Fermi. These discoveries confirmed that $^{238}$U can suffer as many as 17 successive neutron captures and subsequent $\beta^-$ decays if the neutron flux is large enough.

Mendelevium (Md) was produced in 1955 by irradiation of $^{253}$Es with α particles. The name mendelevium was suggested for the element, in honor of the Russian chemist, Dmitri Mendeleev.

In 1958, Ghiorso et al. (1958) announced the positive identification of $^{254}$No that was supposedly produced in the following reaction by use of the double recoil technique

$$^{246}Cm(^{12}C, 4n)^{254}No \xrightarrow[T_{1/2}\approx 3\ s]{\alpha} {}^{250}Fm \xrightarrow[30\ min]{\alpha} \tag{5.7}$$

The first identification of an isotope of element 103 was by the Berkeley group in 1961. The californium isotopes, $^{249, 250, 251, 252}$Cf, were bombarded with boron beams:

$$^{249, 250, 251, 252}Cf(^{10, 11}B, xn)\ ^{258}Lr \xrightarrow[4.3\ s]{\alpha} \tag{5.8}$$

The element was named nobelium after Alfred Nobel.

Lr is the first actinide that was identified through a purely instrumental method, because the half-life of the isotope was too short to allow any chemistry. Silva et al. later conducted the ion-exchange experiments of $^{256}Lr$ and confirmed that Lr exhibits a stable 3+ state in solution, as expected by the actinide concept. Lawrencium was named after Ernest O. Lawrence, the inventor of the cyclotron.

### 词汇

| | |
|---|---|
| artificial radioelement | 人工放射性元素 |
| synthesize | 合成 |
| duration | 持续时间 |
| impurity | 杂质 |
| minor actinide | 次锕系元素 |
| man-made radioelement | 人造放射性元素 |
| periodic table | 元素周期表 |
| manganese | 锰 |
| rhenium | 铼 |
| deuteron | 氘核 |
| cyclotron | 回旋加速器 |
| illinium | 钷(等于 promethium) |
| spectroscopy | 光谱学 |
| infrared | 红外的 |
| emission spectra | 发射光谱 |
| chromatography | 色谱法 |
| yttrium | 钇 |
| lanthanide | 镧系元素 |
| samarium | 钐 |
| trivalent | 三价的 |
| homolog | 同系物 |
| orbital | 轨道 |
| redox | 氧化还原反应 |
| americium | 镅 |
| curium | 锔 |
| europium | 铕 |
| gadolinium | 钆 |
| berkelium | 锫 |
| dysprosium | 镝 |

| | |
|---|---|
| einsteinium | 锿 |
| fermium | 镄 |
| thermonuclear | 热核 |
| coral | 珊瑚 |
| mendelevium | 钔 |
| californium | 锎 |
| nobelium | 锘 |
| lawrencium | 铹 |

### 注释

1. Artificial (also called synthetic or anthropogenic) radioelements are those which have no stable isotopes and do not exist in nature, or at the most only in minute amounts and often as short-lived ephemeral species.

人工(也称为合成或人为)放射性元素是那些没有稳定的同位素,并且不存在于自然界中,或最多只存在微小的数量,并且往往是短寿命核素。

2. In many cases these reactions lead to a mixture of isotopes of a given element for which the composition varies widely with the target, the nature and energy of the projectile, and the duration.

在许多情况下,这些反应会产生某种给定元素的同位素混合物,其成分随靶核、出射核的性质和能量以及持续时间的不同而变化很大。

3. Like any ore, they contain impurities in addition to the desired elements; in this case, the impurities are the fission products other than Pm and Tc together with vast amounts of uninteresting and unwanted material, such as the uranium isotopes.

像任何矿石一样,除了所需的元素外,它们还含有杂质。在这种情况下,杂质是除 Pm 和 Tc 之外的裂变产物,以及大量(研究者)不感兴趣和无用的物质,如铀同位素。

4. Rather ambitious projects are now being elaborated in countries operating reprocessing plants to recover the actinides from U to Am as well as most of the elements formed in the fission process.

运营后处理厂的国家目前正在拟定具有相当挑战性的计划,回收从铀到镅的锕系元素以及裂变过程中形成的大多数元素。

5. On the surface exposed to the deuterons, they found "strong activity, chiefly due to very slow electrons" ascribed to "more than one substance of a half-value period of some months" in addition to $^{32}P$.

在暴露于氘核的表面,他们发现了"主要源自非常缓慢电子的强活度",将其归因于除 $^{32}P$ 之外,"不止一种半衰期为几个月的物质"。

6. This chemistry, ion-exchange chromatography with synthetic resins, was developed during the Manhattan project in order to study the products of nuclear fission, which include several light rare earths.

这种化学方法,即使用合成树脂的离子交换色谱法,是在曼哈顿项目期间开发的,目的是研究核裂变的产物,包括几种轻稀土。

7. By mid-1945, over 30 fission products had been characterized as isotopes of yttrium or one of the lanthanides, but few could be definitely assigned by element and mass numbers.

到 1945 年中期,已经有超过 30 种裂变产物被鉴定为钇或镧系元素的同位素,但很少能通过元素和质量数来确定。

8. According to actinide concept, it was expected that the new series should start with actinium as a prototype and end with the filling of the 5f electron shell at element 103, and that the heavier actinides would be trivalent homologs of the lanthanide series in which 4f orbitals were being filled.

根据锕系的概念,预想新的系列将以锕系元素为开端,以 103 号元素的 5f 电子壳层填充为结束,而更重的锕系将是 4f 轨道被填充的镧系的三价同系物。

9. Element 95, americium, was named after the Americas in analogy with the naming of its rare earth homolog, europium, and curium was named after Marie and Pierre Curie in analogy with the naming of its homolog, gadolinium, that was named after the Finnish rare earth chemist Johann Gadolin.

第 95 号元素镅是以美洲命名的,这与它的稀土同族元素铕的命名相类似;而锔是以玛丽和皮埃尔·居里命名的,这与它的同族元素钆的命名相类似;钆是以芬兰稀土化学家约翰·加多林命名的。

# Chapter 7 Actinide Elements Chemistry

## 7.1 The Actinide Series

The actinide series contains elements with atomic numbers 89 to 103 and is in the seventh period and the third group of the periodic table. The series is the row below the lanthanide series, which is located underneath the main body of the periodic table. The first actinides to be discovered were uranium by Klaproth in 1789 and thorium by Berezelius in 1829, but most of the actinides were man-made products of the 20th century. Actinium and protactinium are found in small portions in nature, as decay products of $^{253}U$ and $^{238}U$. Microscopic amounts of plutonium are made by neutron capture by uranium, and yet occur naturally. Monazite is the principal thorium ore. It is a phosphate ore that contains great amounts of lanthanides in it. The main uranium ore is $U_3O_8$ and is known as pitchblende, because it occurs in black, pitch-like masses. An example of pitchblende is shown in Figure 7.1. Actinide elements all have a high diversity in oxidation numbers and all are radioactive and/or unstable.

**Figure 7.1 A picture of $U_3O_8$, a uranium, pitchblende ore, by Geomartin**

As early as 1923 Bohr suggested that there might exist a group of 15 elements at the end of the periodic table that would be analogous in their properties to the 15 lanthanide (rare earth) elements. This idea, combined with the increasing stability of the +3 oxidation state for the transuranium elements as the atomic number increases from $Z = 93$ to 96, led Seaborg to the conclusion that these new elements constituted a second rare earth series whose initial member was actinium. As the atomic number increases from 90, electrons are added in the 5f subshell similar to the occupation of the 4f subshell in the lanthanides. This series would be terminated with element 103 since this would correspond to the addition of 14 electrons for a completed 5f subshell.

Seaborg's actinide hypothesis was initially a subject of considerable objection since the trivalent oxidation state, unlike in the lanthanide family, was not the most stable in aqueous solution for the elements between $Z = 90$ and 94. In aqueous solution the most stable oxidation states are +4 for thorium, +5 for protactinium, +6 for uranium, +5 for neptunium, and +4 for plutonium. Only for the elements beginning with americium is the +3 the most stable state in solution. Seaborg, however, had correctly identified Ac as the precursor (analogous to La) and Cm as the midpoint element(analogous to Gd). It has shown that mendelevium and nobelium have a divalent state in solution (which probably is the most stable for nobelium). This corresponds to the divalent state observed for ytterbium in the lanthanide elements. For $Z = 90$ to $Z = 94$ the 5f and 6d orbitals are very close in energy and the electronic occupation is variable.

The recognition of the similarity in chemical properties between the actinide and lanthanide elements was an important contributing factor in the synthesis and isolation of the transcurium elements. Most of the chemical identification was carried out by eluting the elements from columns of cation exchange resin. The pattern of the elution behavior from the resin bed of the lanthanide elements made it possible to predict with good accuracy the expected elution position for a new actinide element. This technique constituted the most definitive chemical evidence in the discovery experiments for the elements from atomic numbers 97 to 101.

The electronic ground state of $_{89}Ac$ contains four filled inner electronic shells (the main quantum numbers 1, 2, 3 and 4, designated K, L, M and N, and containing 2, 8, 18 and 32 electrons, respectively). The subsequent outer 5th, 6th and 7th shells (designated "O", "P" and "Q", respectively) are partly empty. In the symbolism used to designate the electronic structure of an atom, the filled shells are usually omitted, as they do not contribute to the chemical properties (and interaction) of the atom. The symbolism $5s^2p^6d^{10}f^0 6s^2p^6d^1 7s^2$ for $_{89}Ac$ indicates that in the 5th shell the subshells s, p, d and f contain 2, 6, 10 and 0 electrons, respectively; the 6th subshell contains 2 s-electrons, 6 p-electrons and 1 d-electron; the outmost 7th shell contains 2 s-electrons. Remembering that the s-, p-, d- and f-subshells can accommodate 2, 6, 10 and 14 electrons respectively, these four outer shells are partly empty. Also leaving out the filled sub-shells, the electronic structure for Ac can be written $5f^0 6d^1 7s^2$. There can be interaction between the 5f, 6d and 7s orbitals, as a consequence of their very similar binding energies. As these partly empty outer orbitals are the ones which may take part in the chemical bondings we can expect the actinides to show a highly variable chemistry.

The main feature of the electronic structure, when going from $_{89}Ac$ to $_{103}Lr$ is the successive filling of the 5f subshell. This leads to a slight contraction of the atomic and ionic radii with increasing atomic number (the actinide contraction). This contraction is due to the successive addition of electrons in an inner f shell where the incomplete screening of the nuclear charge by the added f electron leads to a contraction of the outer valence orbital. However, there are irregularities, which are attributed to the extra stability of the half-filled 5f subshell (at 7 electrons), leading to more labile 6d-electrons.

## 7.2 Actinide Oxidation States

The electrons in the 6d and 7s subshells are more loosely bound than the electrons in the filled subshells, in general, also more loosely than the 5f electrons. In these outer shells the binding energies are on the range of a few eV, i. e., the same order of magnitude as is common in chemical bonding. Thus, it is understandable that Ac easily looses its $6d7s^2$ electrons to form $Ac^{3+}$, and Th its $6d^27s^2$ electrons to form $Th^{4+}$. For the subsequent elements, from Pa to Am, the situation is more complicated. There are reasons to suppose that the spatial characteristics of the f-subshell orbitals may change abruptly at certain atomic numbers; that is, the f-shell electrons may be shielded more strongly in some elements than in others where the f-orbitals extend close to the surface of the electronic cloud (where chemical interaction occurs), and where the 5f-electrons are in closer contact with the d- and s-shell electrons. There can be little doubt that the 5f-electrons are present in all of the actinide elements after Pa.

Most actinides exhibit a larger variety of valence states, and the most stable are +6 for uranium, +5 for protactinium and neptunium, +4 for thorium and plutonium and +3 for actinium and other actinides. In solution, the 2+, 3+, and 4+ species are present as metal cations, while the higher oxidation states are present as oxo-cations, $MO_2^+$ and $MO_2^{2+}$. The known oxidation states of the actinide elements are shown in Table 7. 1.

**Table 7. 1 Oxidation states of the actinide elements**

| Atomic Number: | 89 | 90 | 91 | 92 | 93 | 94 | 95 | 96 | 97 | 98 | 99 | 100 | 101 | 102 | 103 |
|---|---|---|---|---|---|---|---|---|---|---|---|---|---|---|---|
| Element: | Ac | Th | Pa | U | Np | Pu | Am | Cm | Bk | Cf | Es | Fm | Md | No | Lr |
| Oxidation States | | | | | | | | | | | | | | | |
| | | | | | | | | | | | | | 1? | | |
| | | | | | | | (2) | (2) | | (2) | (2) | 2 | $\underline{2}$ | $\underline{2}$ | |
| | $\underline{3}$ | (3) | (3) | 3 | 3 | 3 | $\underline{3}$ | $\underline{3}$ | $\underline{3}$ | $\underline{3}$ | $\underline{3}$ | $\underline{3}$ | $\underline{3}$ | $\underline{3}$ | $\underline{3}$ |
| | | $\underline{4}$ | 4 | 4 | 4 | $\underline{4}$ | 4 | 4 | 4 | 4 | (4) | 4? | | | |
| | | | $\underline{5}$ | 5 | $\underline{5}$ | 5 | 5 | 5? | | 5? | | | | | |
| | | | | $\underline{6}$ | 6 | 6 | 6 | 6? | | | | | | | |
| | | | | | 7 | (7) | 7? | | | | | | | | |

The most common oxidation states are underlined, unstable oxidation states are shown in parentheses.

Question marks indicate species that have been claimed but not substantiated.

Source: From Katz, et al. (1980).

The pentavalent state of the actinides (except for Pa and Np) is less stable than the other states and normally undergoes disproportionation in acid solutions. Plutonium is particularly interesting in the variety of oxidation states that can coexist in aqueous solutions. For example, a plutonium solution in 0. 5 M HCl of $3\times10^{-4}$ M Pu concentration at 25 ℃, which is initially 50%

Pu(Ⅳ) and 50% Pu(Ⅵ), will equilibrate within a few days via disproportionation reactions to an equilibrium system that is 75% Pu(Ⅵ), 20% Pu(Ⅳ), and few percent each of Pu(Ⅴ) and Pu(Ⅲ), in the absence of complexing anions. The reactions are

$$2PuO_2^+ + 4H^+ \longleftrightarrow Pu^{4+} + PuO_2^{2+} + 2H_2O \quad (7.1)$$

and

$$PuO_2^+ + Pu^{4+} \longleftrightarrow PuO_2^{2+} + Pu^{3+} \quad (7.2)$$

Although Pu(Ⅴ) readily disproportionates at concentrations $\geqslant 10^{-8}$ M in acidic solutions, it is the state observed in more basic natural and ocean waters, partly due to the stability of the bicarbonate complex.

## 7.3 Actinide Complexes

Since the differences in energy of the electronic levels are similar to chemical bond energies, the most stable oxidation of the actinides may change from one chemical compound to another, and the solution chemistry will be sensitive to the ligands present. Thus complex formation becomes an important feature of the actinide chemistry.

The chemical properties are different for the different valency states, while in the same valency state the actinides closely resemble each other. These properties have extensively exploited for the separation and isolation of the individual elements in pure form.

The compounds formed are normally quite ionic. The ionic radii of the actinide elements of the different valency states decreases with increasing atomic number. Consequently the charge density of the actinide ions increase with increasing atomic number and, therefore, the probability of formation of complexes and of hydrolysis increases with atomic number.

The pattern of stabilities of complexes in the tetravalent states and hexavalent states is the same and follows the order of decreasing ionic radius,

$$Th^{4+} < U^{4+} < Np^{4+} < Pu^{4+} \quad (7.3)$$

$$UO_2^{2+} < NpO_2^{2+} < PuO_2^{2+} \quad (7.4)$$

This also explains that the extraction behavior for the $M^{4+}$ ions, $Pu^{4+}$ is better extracted than $Th^{4+}$. In the case of tributyl phosphate, TBP, dissolved in kerosene, the extracted M(Ⅵ), M(Ⅳ) and M(Ⅲ) species are $MO_2(NO_3)_2(TBP)_2$, $M(NO_3)_4(TBP)_2$, and $M(NO_3)_3(TBP)_3$, respectively.

For the same element, the stability of the complexes varies with the oxidation state in the series

$$M^{4+} \geqslant MO_2^{2+} > M^{3+} \geqslant MO_2^+ \quad (7.5)$$

The reversal between $M^{+3}$ and $MO_2^{2+}$ reflects that the hexavalent metal atom in the linear $[OMO]^{2+}$ is only partially shielded by the two oxygen atoms; thus the metal ion $MO_2^{2+}$ has a higher charge density than $M^{3+}$ (i. e., about 3.2 ± 0.1). Similarly in $MO_2^+$, the effective charge is about 2.2±0.1. Of importance in reprocessing are the low distribution ratios of fission products like Cs, Sr, Ru and Zr.

The extraction of trivalent actinides in general follows the sequence

$$Ac^{3+} < Am^{3+} < Cm^{3+} \text{ etc} \tag{7.6}$$

for example, $Cm^{3+}$ is extracted easier, i. e., at a lower pH, than $Am^{3+}$, as the ionic radii decrease in that order.

Hydrolysis is actually a special type of complex ion formation. The large positive charge associated with transuranium cations that leads to hydrolysis is also the driving force for the interaction of nucleophiles with the transuranium cations. Water is only one example of a nucleophilic ligand. Other nucleophilic ligands present in solution may replace water molecules directly bound to the metal cation to form inner sphere complexes or alternatively, they may displace water molecules only from the outer hydrate shell to form outer sphere complexes.

The actinide cations are "hard acids", that is, their binding to ligands is described in terms of electrostatic interactions, and they prefer to interact with hard bases such as oxygen or fluorine rather than softer bases such as nitrogen or sulfur. The actinide cations do form complexes with the soft bases but only in nonaqueous solvents. As typical hard acids, the stabilities of the actinide complexes are due to favorable entropy effects. The enthalpy terms are either endothermic or very weakly exothermic and are of little importance in determining the overall position of the equilibrium in complex formation. Although there is some variation within the given cation types, the general order of complexing power of different anions is $F^- > NO_3^- > Cl^- > ClO_4^-$ for singly charged anions and $CO_2^{2-} > C_2O_4^{2-} > SO_4^{2-}$ for doubly charged anions.

## 7.4 Chemical Properties of Actinides

Like the lanthanides, all actinides are highly reactive with halogens and chalcogens; however, the actinides react more easily.

Actinium is chemically similar to lanthanum, which is explained by their similar ionic radii and electronic structures. Like lanthanum, actinium almost always has an oxidation state of +3 in compounds, but it is less reactive and has more pronounced basic properties. Among other trivalent actinides $Ac^{3+}$ is least acidic, i. e. has the weakest tendency to hydrolyze in aqueous solutions.

Thorium is rather active chemically. Owing to lack of electrons on 6d and 5f orbitals, the tetravalent thorium compounds are colorless. At pH < 3, the solutions of thorium salts are dominated by the cations $[Th(H_2O)_8]^{4+}$. The $Th^{4+}$ ion is relatively large, and depending on the coordination number can have a radius between 0.95 and 1.14 Å. As a result, thorium salts have a weak tendency to hydrolyze. The distinctive ability of thorium salts is their high solubility both in water and polar organic solvents.

Protactinium exhibits two valence states: the +5 is stable, and the +4 state easily oxidizes to protactinium(Ⅴ). Thus tetravalent protactinium in solutions is obtained by the action of strong reducing agents in a hydrogen atmosphere. Tetravalent protactinium is chemically similar to uranium(Ⅳ) and thorium(Ⅳ). Fluorides, phosphates, hypophosphate, iodate and phenyl arsonates of protactinium(Ⅳ) are insoluble in water and dilute acids. Protactinium forms soluble carbonates. The hydrolytic properties of pentavalent protactinium are close to those of tantalum

(Ⅴ) and niobium(Ⅴ). The complex chemical behavior of protactinium is a consequence of the start of the filling of the 5f shell in this element.

Uranium has a valence from 3 to 6, the last being most stable. In the hexavalent state, uranium is very similar to the group 6 elements. Many compounds of uranium(Ⅳ) and uranium (Ⅵ) are non-stoichiometric, i. e. have variable composition. For example, the actual chemical formula of uranium dioxide is $UO_{2+x}$, where x varies between 0. 4 and 0. 32. Uranium (Ⅵ) compounds are weak oxidants. Most of them contain the linear "uranyl" group, $UO_2^{2+}$. Between 4 and 6 ligands can be accommodated in an equatorial plane perpendicular to the uranyl group. The uranyl group acts as a hard acid and forms stronger complexes with oxygen-donor ligands than with nitrogen-donor ligands. $NpO_2^{2+}$ and $PuO_2^{2+}$ are also the common form of Np and Pu in the +6 oxidation state. Uranium (Ⅳ) compounds exhibit reducing properties, e. g., they are easily oxidized by atmospheric oxygen. Uranium (Ⅲ) is a very strong reducing agent. Owing to the presence of d-shell, uranium (as well as many other actinides) forms organometallic compounds, such as $U^{III}(C_5H_5)_3$ and $U^{IV}(C_5H_5)_4$.

Neptunium has valence states from 3 to 7, which can be simultaneously observed in solutions. The most stable state in solution is +5, but the valence +4 is preferred in solid neptunium compounds. Neptunium metal is very reactive. Ions of neptunium are prone to hydrolysis and formation of coordination compounds.

Plutonium also exhibits valence states between 3 and 7 inclusive, and thus is chemically similar to neptunium and uranium. It is highly reactive, and quickly forms an oxide film in air. Plutonium reacts with hydrogen even at temperatures as low as 25-50 ℃; it also easily forms halides and intermetallic compounds. Hydrolysis reactions of plutonium ions of different oxidation states are quite diverse. Plutonium(Ⅴ) can enter polymerization reactions.

The largest chemical diversity among actinides is observed in americium, which can have valence between 2 and 6. Divalent americium is obtained only in dry compounds and non-aqueous solutions (acetonitrile). Oxidation states +3, +5 and +6 are typical for aqueous solutions, but also in the solid state. Tetravalent americium forms stable solid compounds (dioxide, fluoride and hydroxide) as well as complexes in aqueous solutions. It was reported that in alkaline solution americium can be oxidized to the heptavalent state, but these data proved erroneous. The most stable valence of americium is 3 in the aqueous solutions and 3 or 4 in solid compounds.

Valence 3 is dominant in all subsequent elements up to lawrencium (with the exception of nobelium). Curium can be tetravalent in solids (fluoride, dioxide). Berkelium, along with a valence of +3, also shows the valence of +4, more stable than that of curium; the valence 4 is observed in solid fluoride and dioxide. The stability of $Bk^{4+}$ in aqueous solution is close to that of $Ce^{4+}$. Only valence 3 was observed for californium, einsteinium and fermium. The divalent state is proved for mendelevium and nobelium, and in nobelium it is more stable than the trivalent state. Lawrencium shows valence 3 both in solutions and solids.

### 词汇

actinide series　　锕系

| | |
|---|---|
| seventh period | 第七周期 |
| third group | 第三族 |
| lanthanide series | 镧系 |
| rare earth metal | 稀土金属 |
| microscopic amount | 微量 |
| actinium | 锕 |
| plutonium | 钚 |
| crystal | 晶体 |
| transition metal | 过渡金属 |
| variable valence | 可变化合价 |
| ionization chamber | 电离室 |
| smoke detector | 烟雾探测器 |
| transient | 瞬态 |
| debris | 碎片 |
| f-block element | f 区元素 |
| halogen | 卤素 |
| chalcogen | 硫族 |
| hydrolyze | 水解 |
| tetravalent | 四价的 |
| solubility | 溶解度 |
| non-stoichiometric | 非化学计量 |
| hard acid | 硬酸 |
| organometallic compounds | 有机金属化合物 |
| reactive | 反应的 |
| halide | 卤化物 |
| intermetallic compound | 金属间化合物 |
| polymerization reaction | 聚合反应 |
| redox potential | 氧化还原电位 |
| alkaline solution | 碱性溶液 |

**注释**

1. The actinide series contains elements with atomic numbers 89 to 103 and is in the seventh period and the third group of the periodic table.

锕系元素是原子序数为 89~103 的元素，属于元素周期表的第七周期和第三族。

2. Actinide elements all have a high diversity in oxidation numbers and all are radioactive and/or unstable.

锕系元素的氧化态数目具有很多种，而且锕系元素都具有放射性和/或不稳定性。

3. The electrons in the 6d and 7s subshells are more loosely bound than the electrons in the filled subshells, in general, also than the 5f electrons. In these outer shells the binding energies are on the range of a few eV, i. e., the same order of magnitude as is common in chemical

bonding.

6d 和 7s 亚壳层的电子比填满的亚壳层中的电子束缚更松散,通常也比 5f 的电子束缚更松散。在这些外壳层中,结合能一般为几个电子伏特,与化学键的键能处于同一数量级。

4. Owing to lack of electrons on 6d and 5f orbitals, the tetravalent thorium compounds are colorless.

四价钍由于在 6d 和 5f 轨道上缺少电子,故其化合物是无色的。

5. Tetravalent protactinium is chemically similar to uranium(Ⅳ) and thorium(Ⅳ).

四价镤与铀(Ⅳ)和钍(Ⅳ)的化学性质相似。

6. Hydrolysis reactions of plutonium ions of different oxidation states are quite diverse. Plutonium(Ⅴ) can enter polymerization reactions.

不同氧化态的钚离子的水解反应多种多样。钚(Ⅴ)可以发生聚合反应。

# Chapter 8 Solvent Extraction and Ion-Exchange Separation Techniques

## 8.1 Separation Techniques

The task of quantitative and effective separation of small amounts of radionuclides has appreciably enhanced the development of modern separation techniques. High radionuclide purity is of great importance for applications in nuclear medicine as well as for sensitive measurements. In this context, impurities of long-lived radionuclides are of particular importance, because their relative activity increases with time. For example, if the activity of $^{90}Sr$ is only 0.1% of that of $^{140}Ba$ after fresh separation, it will increase to 11.5% in three months. The most frequently used separation techniques are crystallization, precipitation, or coprecipitation, electrolysis, distillation, solvent extraction, ion-exchange, and chromatography.

Precipitation is only applicable if the solubility product is exceeded, that is, if the concentration of the radionuclide to be separated is high enough. If the concentration is too low, coprecipitation may be applied by addition of a suitable carrier, such as stable compounds of identical or very similar chemical properties. Application of isotopic carriers is very effective, but it leads to a decrease in specific activity. Therefore, non-isotopic carriers with suitable chemical properties are preferred. Hydroxides such as iron (Ⅲ) hydroxide or other sparingly soluble hydroxides give high coprecipitation yields, because of their high sorption capacity. Coprecipitation of actinides with $LaF_3$ is an important example.

Electrolytic deposition of radionuclides is frequently applied. It gives thin samples and is well suited for preparation of standard samples. For instance, Po, Pb, or Mn can be deposited with high yields on anodes of Cu, Pt, or Ag, and by the electrolysis of the nitrates or chlorides of Th and Ac in acetone or ethanol solutions these elements can be separated on cathodes. The preparation of thin samples by electrolytic deposition is of special interest for the measurement of $\alpha$ emitters, such as isotopes of Pu or other actinides.

Separation of radionuclides by distillation is applicable if volatile compounds are formed. $^{131}I$ can be separated from irradiated Te by distillation. Other examples are separation of Ru as $RuO_4$ under oxidizing conditions, and volatilization of Tc as $Tc_2O_7$ from concentrated $H_2SO_4$ at 150 to 250 ℃. $^{32}P$ may be purified by volatilization as $PCl_5$ in a stream of $Cl_2$.

Solvent extraction is widely used for separation of radionuclides, because this technique is simple, fast, and applicable in the range of low concentrations. Addition of a carrier is not required. Solvent extraction plays an important role in reprocessing, and tributyl phosphate (TBP) is a preferred complexing agent for the separation and purification of U and Pu.

Ion-exchange procedures have also found broad application in radiochemistry. Commercial ion-exchange resins exhibit relatively low selectivities. Higher values of selectivity are obtained by use of organic ion exchangers carrying chelating groups of high selectivity as anchor groups or by application of inorganic ion exchangers. Highly selective organic ion exchangers are synthesized on the basis of polystyrene, cellulose, or other substances as matrices. High selectivity with commercial ion-exchange resins is also obtained by the addition of complexing agents, such as α-hydroxycarboxylic acids to the solution.

Chromatographic separation techniques are based on adsorption, ion-exchange, or partition between a stationary and a mobile phase. Gas chromatography (GC) is applied for the separation of volatile compounds. Thermochromatography (using a temperature gradient) or isothermal gas chromatography is frequently used for the study of the properties of small amounts of radionuclides.

## 8.2 Solvent Extraction Theory

The definition of solvent extraction is "the process of transferring a substance from any matrix to an appropriate liquid phase". If the substance is initially present as a solute in an immiscible liquid phase, the process is synonymous with liquid-liquid extraction.

Most inorganic species are only slightly soluble in inert organic solvents. Consequently, the extraction can occur only through interaction with an organic molecule, the extractant. The organic phase sometimes consists of a pure extractant, but more commonly, the extractant is dissolved in an inert organic diluent. In industrial extraction, the diluent is often a hydrocarbon-like kerosene or toluene. In small-scale extraction a variety of diluents is used, e. g., chloroform, carbon tetrachloride, and many others. The extractant is defined as "the active component(s) primarily responsible for transfer of a solute from one phase to the other", but usually the word denotes all components of the solvent that are involved in the extraction. Usually the term "extraction" is reserved for the transfer from an aqueous to an organic phase, while the opposite transfer direction is called strip or back-extraction.

Solvent extraction is based on the distribution of a solute between two immiscible phases. Gibbs' phase rule can be applied, namely:

$$P + F = C + 2 \tag{8.1}$$

where $P$ is the number of phases, $F$ is the number of degrees of freedom, and $C$ is the number of components. Thus, a system consisting of two immiscible solvents and one solute distributed between them has one degree of freedom at constant temperature and pressure. Consequently, if the solute concentration in one phase is constant, the concentration in the other phase is also constant. The relationship between these concentrations led to Nernst's distribution law. This law can be expressed as

$$k_d = \frac{c_{org}}{c_{aq}} \tag{8.2}$$

where $c_{org}$ is the concentration of an extractable species in the organic phase and $c_{aq}$ is the concentration in the aqueous phase. Thus, $k_d$, referred to as the distribution factor, is dimensionless. From Eq. (8.2) it follows that the distribution factor does not depend on the total solute concentration or the phase volumes.

The expression given for $k_d$ is valid if there is only one species that distributes between the phases. Usually the situation is more complicated; e.g., in metal ion-extraction, the metal can be present in the aqueous phase as free metal ion as well as positive, neutral, or negative complexes. In the organic phase, it is present as an uncharged complex. For such cases, the distribution ratio is used. It is abbreviated $D$ and it is defined as

$$D = \frac{c_{org}}{c_{aq}} \tag{8.3}$$

where $c_{org}$ is the total analytical concentration of, e.g., a metal in the organic phase and $c_{aq}$ is the corresponding total analytical concentration in the aqueous phase. $D$ is easy to determine by radiometric methods since it is equal to the ratio between the specific activities of the phases. This is one reason why liquid-liquid extraction is a common method in radiochemistry.

The extracted fraction of an element can be calculated using the expression

$$\%E = \frac{DR}{(1 + DR)} \times 100 \tag{8.4}$$

$R$ is the phase ratio, that is, the volume ratio of organic phase to the aqueous phase. Another useful definition is that of the separation factor, $SF_{AB}$, between the elements A and B in a liquid-liquid system. It is simply defined as

$$SF_{AB} = \frac{D_A}{D_B} \tag{8.5}$$

## 8.3 Extraction Systems

There are four major types of extraction systems that can be used for separation procedures.

Extraction of simple inorganic compounds can be exemplified by the classical distribution of iodine between water and carbon tetrachloride or by the extraction of, e.g., $RuO_4$, $OsO_4$, or $AsI_3$ into an organic solvent. Such extraction procedures are usually rapid, selective, and are well suited for radiochemical separations.

A metal ion, $M^{n+}$, may also form a complex with an organic anion $A^-$ to form a neutral, hydrophobic complex $MA_n$. The organic reagent is often an acid, HA, and the anion $A^-$ a bidentate chelating agent. The mechanism is therefore often referred to as chelate extraction.

Another possibility to extract a metal ion is to replace its hydrate water by a solvating reagent. This makes the metal species less hydrophilic and it can be transferred to an organic phase. The solvating reagent can be the organic phase or it can be dissolved in a diluent. A well-known example is the extraction of uranium by tri-n-butyl phosphate from nitric acid:

$$UO_2^{2+} + 2NO_3^- + 2\overline{TBP} \leftrightarrow \overline{UO_2(NO_3)_2(TBP)_2} \tag{8.6}$$

In ion pair extraction the concentration of the inorganic anion $L^-$ is high enough to form the complex $ML_n^{(\nu-n)-}$. This complex usually is an inner sphere complex that is more weakly hydrolyzed than the uncomplexed metal ion. It can therefore be extracted as an ion pair with the organic cation $Q^+$. Metal anions like $TcO_4^-$ and $ReO_4^-$ can, of course, be extracted without any extra inorganic anions. Q often is a primary, secondary, or tertiary amine or a quaternary ammonium salt. These types of extractants are often referred to as liquid ion exchangers because of the similarity between the ion-pair formation process and an ion-exchange process. Negatively charged ion-pair extractants (e. g., sulfonic acids) that directly extract metal ions also exist. They are not, however, of great importance because of their poor selectivity.

There are also other types of extractants such as crown ethers that have a ring structure that fits an atom with a certain ionic radius, thus forming a hydrophobic and extractable species. These reagents have been used, e. g., for isolation of cesium and strontium. Many extractants also combine several of the mechanisms mentioned above. Sometimes a mixture of two extractants gives a higher distribution ratio than the extractants separately. This is referred to as synergy. The reason for synergy is that one extractant neutralizes the charge of the metal ion while the other replaces the remaining hydrate water molecules.

## 8.4 Extraction Equipment

The solvent extraction equipment utilized for industrial-scale aqueous reprocessing must enable continuous processing at high throughputs while efficiently mixing and separating the two phases. In the nuclear industry, specific constrains, such as remote operation and maintenance must be considered, since the solutions processed are highly radioactive. There are three basic types of equipment used in industrial-scale nuclear solvent extraction processes: mixer-settlers, columns and centrifugal contactors.

Mixer-settlers consist of a mixing chamber, usually with an impeller that also pumps the liquid through the system, and a separation chamber in which the phases are separated by gravity. The size ranges from laboratory equipment with a volume of a few milliliters per step to industrial settlers with a volume of several thousand cubic meters. The main advantages of mixer-settlers are that they are easy to construct, cheap, easy to run, need very little service, have well-defined steps, and also that it is easy to add more steps to an existing facility. The disadvantages are that they contain a large organic phase volume (= high capital costs) and also that it takes a rather long time to reach steady-state after start-up of the plant. On the other hand, steady-state is not destroyed if halting the system, i. e., restart is very simple. In the nuclear industry, the large volumes make mixer-settlers less suitable for elements that may cause criticality problems.

Columns are either dynamic (e. g., pulse columns) or static, i. e., without mechanical mixing of the phases. The size and capacity cover a wide range, from a few centimeters diameter and about 50 cm length to 4 m diameter and 60 m height. The main advantage of columns is that a large number of steps can be contained in one column. The disadvantages are that they are more difficult to design and scale-up than mixer-settlers and also that they are sensitive to load variations and crud (e. g., solid particles in the solutions). They cannot be halted without losing steady-state and it is also impossible to add more steps once the column has been constructed. An advantage in the nuclear industry is that it is easy to construct columns that are safe against criticality (the maximum volume is well defined) provided that the concentration of fissile material is kept within the design limits.

Centrifugal contactors are used when rapid phase disengagement is necessary to prevent chemical or radiolytical degradation of an extractant or an extracted compound. They are also an attractive alternative if the requirements on the phase purity are high. The liquid holdup volumes are often small, and therefore this type of equipment is also suitable if expensive reagents are used. The difference between a centrifuge and a centrifugal contactor is mainly that the centrifuge has an external mixer and each centrifuge is one well-defined step, while a centrifugal extractor has internal mixing and several steps integrated within one unit. The advantages of centrifugal contactors are that they have high capacity despite their rather small volumes and that they give a good and rapid phase separation. The disadvantages are that they are mechanically complicated, difficult to scale up or down and sensitive to solid entrainment. Centrifuges are used for nuclear science but not very much in the nuclear industry. This may change in the future, when partitioning and transmutation processes may require extraction with short contact times to avoid problems with radiolysis.

## 8.5 Ion-Exchange Theory

Ion-exchange resins are either cation or anion exchangers. The former group may be classified as strong, moderate, or weak acid cation exchangers, while the anion exchangers are classified into weak and strong base exchangers. There are also other types of ion exchangers, e. g., chelating resins and bifunctional resins. A typical ion exchanger consists of functional groups attached to a polymer matrix. As an example, a strong acid type cation exchanger typically consists of sulfonic acid groups bound to a cross-linked polystyrene matrix. The exchange reaction can be written as:

$$R - SO_3^- H^+ + M^+ X^- \leftrightarrow R - SO_3^- M^+ + H^+ X^- \tag{8.7}$$

where $M^+$ is a metal ion and $X^-$ is an anionic ligand.

Moderately strong cation exchangers contain-$PO(OH)_2$ as a functional group, while weak acid cation exchangers contain the functional group-COOH. Strong base anion exchangers contain the group-$(CH_2NR_3)Cl$, while weak base anion exchangers contain-$(CH_2NHR_2)Cl$ or-$(CH_2NH_2R)$ Cl. In these formulas, R represents an alkyl chain.

For the exchange reaction given in Eq. (8.7) a selectivity coefficient can be defined as

$$K_s \frac{[R - SO_3M] \cdot [HX]}{[R - SO_3H] \cdot [MX]} \tag{8.8}$$

The selectivity coefficient depends on several parameters, e. g., on the type of ion exchanger, the charge of the exchanged ion, and the conditions of the solution. It is also possible to define a distribution ratio for the ion exchanger. This ratio is expressed as

$$K_D = c_{res}/c_{aq} \tag{8.9}$$

where $c_{res}$ is the equilibrium concentration of the exchanged ion in the resin (mol/kg) and $c_{aq}$ is the concentration in the aqueous solution ($mol/m^3$).

Ion exchangers are always solid and operated as columns, membranes, or surfaces covered with ion exchanger. The size of such columns ranges from a few millimeters up to several cubic meters, and the flow capacities from a few milliliters per minute to several cubic meters per hour.

**词汇**

| | |
|---|---|
| coprecipitation | 共沉淀 |
| electrolysis | 电解 |
| distillation | 蒸馏 |
| solvent extraction | 溶剂萃取 |
| specific activity | 比活度 |
| complexing agent | 络合剂 |
| commercial ion-exchange resin | 商业离子交换树脂 |
| chelating group | 螯合基团 |
| anchor group | 锚定基团 |
| immiscible | 不互溶的 |
| kerosene | 煤油 |
| toluene | 甲苯 |
| chloroform | 氯仿 |
| carbon tetrachloride | 四氯化碳 |
| extractant | 萃取剂 |
| hydrophilic | 亲水性的 |
| hydrophobic | 疏水性的 |
| solvating reagent | 溶剂化试剂 |
| bidentate chelating agent | 双齿螯合剂 |
| neutral complex | 中性络合剂 |
| synergistic effect | 协同效应 |
| throughput | 产量 |
| impeller | 推动器 |
| mixer-settler | 混合澄清槽 |
| centrifugal contactor | 离心萃取器 |
| sulfonic acid groups | 磺酸基团 |
| alkyl chain | 烷基链 |
| anion | 阴离子 |
| cation | 阳离子 |
| polystyrene | 聚苯乙烯 |

**注释**

1. Solvent extraction plays an important role in reprocessing, and tributyl phosphate (TBP) is a preferred complexing agent for the separation and purification of U and Pu.

溶剂萃取在后处理中发挥着重要作用,磷酸三丁酯(TBP)是分离和纯化 U 和 Pu 的首选络合剂。

2. Higher values of selectivity are obtained by use of organic ion exchangers carrying chelating groups of high selectivity as anchor groups or by application of inorganic ion exchangers.

通过使用携带高选择性螯合基团作为锚定基团的有机离子交换剂,或通过应用无机离子交换剂可获得更高的选择性值。

3. The definition of solvent extraction is "the process of transferring a substance from any matrix to an appropriate liquid phase". If the substance is initially present as a solute in an immiscible liquid phase, the process is synonymous with liquid-liquid extraction.

溶剂萃取的定义是"将物质从任何基质转移到适当的液相的过程"。如果物质最初存在于不相溶的液相中作为溶质,则该过程与液-液萃取同义。

4. The extractant is defined as "the active component(s) primarily responsible for transfer of a solute from one phase to the other", but usually the word denotes all components of the solvent that are involved in the extraction. Usually the term "extraction" is reserved for the transfer from an aqueous to an organic phase, while the opposite transfer direction is called strip or back-extraction.

萃取剂被定义为"主要负责将溶质从一相转移到另一相的活性组分",但通常该词表示参与萃取的溶剂的所有组分。通常,术语"萃取"是指从水相转移到有机相,而相反的转移方向则被称为反萃取。

5. Sometimes a mixture of two extractants gives a higher distribution ratio than the extractants separately. This is referred to as synergy.

有时,两种萃取剂的混合物比其中任何一种单独的萃取剂具有更高的分配比,这被称为协同作用。

6. The solvent extraction equipment utilized for industrial-scale aqueous reprocessing must enable continuous processing at high throughputs while efficiently mixing and separating the two phases.

用于工业规模的水性后处理的溶剂提取设备必须能够在高产量下进行连续处理,同时可有效地混合和分离两相。

7. Centrifuges or centrifugal contactors are used when rapid phase disengagement is necessary to prevent chemical or radiolytical degradation of an extractant or an extracted compound.

当需要快速相分离以防止萃取剂或萃取化合物发生化学或放射性降解时,需要使用离心机或离心萃取器。

8. The former group may be classified as strong, moderate, or weak acid cation exchangers, while the anion exchangers are classified into weak and strong base exchangers.

前者(阳离子交换剂)可分为强酸、中强酸或弱酸阳离子交换剂,而阴离子交换剂可分为强碱和弱碱交换剂。

# Chapter 9 Uranium Mining

## 9. 1 Uranium Resources

Uranium is found throughout the Earth's crust in various concentrations with an average abundance of approximately 2. 8 g/t. To put this value into perspective, uranium is more abundant than gold but less than copper. Australia, Canada, and Kazakhstan have the highest RAR of uranium, whereby Australia has 30% in total identified resources in the< 130 USD/kgU price category. A regularly updated report on uranium resources is jointly published by the Organization for Economic Co-operation and Development (OECD) Nuclear Energy Agency (NEA) and the International Atomic Energy Agency (IAEA). The global distribution of identified uranium resources is in the< 130 USD/kgU price category. The distribution of RAR and inferred resources varies not only by nation state but also by price category. The production cost of uranium depends on many factors, including the method used to mine the ore, geography, geology, the quality of grade of the ore, and the mineral type.

## 9. 2 Mining Methods

The mining process involves the extraction of raw uranium bearing minerals out of the ground, which is accomplished with various techniques. In 2016, the percentage of the mined uranium produced by each mining method was: in-situ leaching (49. 7%), underground mining (30. 8%), open pit (12. 9%), heap leaching (0. 4%), co-product/by-product (6. 1%). The remaining 0. 1% was derived as miscellaneous recovery.

A generic schematic of the acid-leaching process is shown in Figure 9. 1.

Open pit: In open pit mining, overburden is removed by drilling and blasting to expose the ore body, which is then mined by blasting and excavation using loaders and dump trucks. Workers spend much time in enclosed cabins thus limiting exposure to radiation. Water is extensively used to suppress airborne dust levels. Groundwater is an issue in all types of mining, but in open pit mining, the usual way of dealing with it — i. e. when the target mineral is found below the natural water table — is to lower the water table by pumping off the water. The ground may settle considerably when groundwater is removed and may again move unpredictably when groundwater is allowed to rise again after mining is concluded. Land reclamation after mining takes different routes, depending on the amount of material removed. Due to the high energy density of uranium, it is often sufficient to fill in the former mine with the overburden, but in case of a mass deficit exceeding the height difference between the previous surface level and the natural water table, artificial lakes develop when groundwater removal ceases. If sulfites, sulfides or

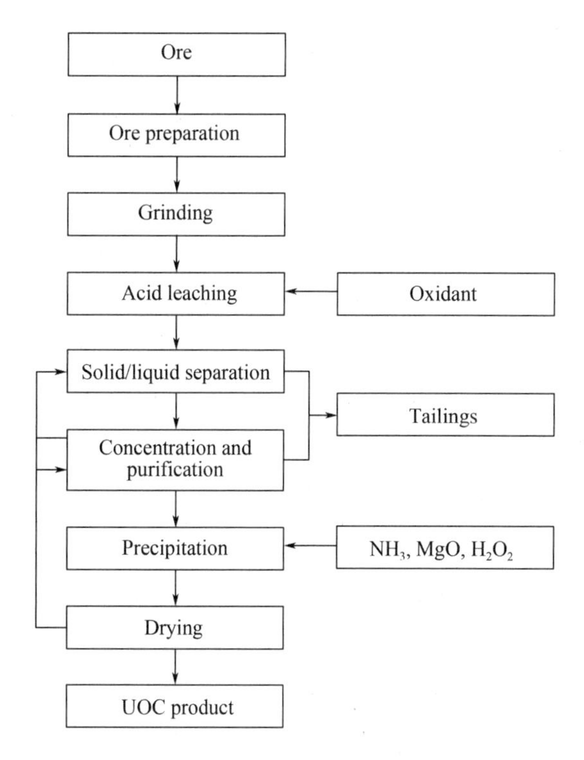


**Figure 9.1 A generic schematic of the acid-leaching process**

sulfates are present in the now-exposed rocks acid, mine drainage can be a concern for those newly developing bodies of water. Mining companies are now required by law to establish a fund for future reclamation while mining is ongoing and those funds are usually deposited in such a way as to be unaffected by bankruptcy of the mining company.

Underground: If the uranium is too far below the surface for open pit mining, an underground mine might be used with tunnels and shafts dug to access and remove uranium ore. If the uranium is too far below the surface for open pit mining, an underground mine might be used with tunnels and shafts dug to access and remove uranium ore. The stope, which is the workshop of the mine, is the excavation from which the ore is extracted. Three methods of stope mining are commonly used. In the "cut and fill" or "open stoping" method, the space remaining following removal of ore after blasting is filled with waste rock and cement. In the "shrinkage" method, only sufficient broken ore is removed via the chutes below to allow miners working from the top of the pile to drill and blast the next layer to be broken off, eventually leaving a large hole. The method known as "room and pillar" is used for thinner, flatter ore bodies. In this method, the ore body is first divided into blocks by intersecting drives, removing ore while so doing, and then systematically removing the blocks, leaving enough ore for roof support. The health effects discovered from radon exposure in unventilated uranium mining prompted the switch away from uranium mining via tunnel mining towards open cut and in-situ leaching technology, a method of extraction that does not produce the same occupational hazards, or mine tailings, as conventional mining. With regulations in place to ensure the use of high volume ventilation technology if any confined space uranium mining is occurring, occupational exposure and mining deaths can be largely eliminated. The Olympic Dam and Canadian underground mines are ventilated with powerful fans with radon levels being kept at a very low to practically "safe level" in uranium mines. Naturally occurring radon in other, non-uranium mines, also may need control by ventilation.

Heap leaching: Heap leaching is an extraction process by which chemicals (usually sulfuric acid) are used to extract the economic element from ore which has been mined and placed in piles on the surface. Heap leaching is generally economically feasible only for oxide ore deposits. Oxidation of sulfide deposits occurs during the geological process called weathering. Therefore, oxide ore deposits are typically found close to the surface. If there are no other economic elements within the ore a mine might choose to extract the uranium using a leaching agent, usually a low molar sulfuric acid. If the economic and geological conditions are right, the mining company will level large areas of land with a small gradient, layering it with thick plastic (usually HDPE or LLDPE), sometimes with clay, silt or sand beneath the plastic liner. The extracted ore will typically be run through a crusher and placed in heaps atop the plastic. The leaching agent will then be sprayed on the ore for 30 – 90 d. As the leaching agent filters through the heap, the uranium will break its bonds with the oxide rock and enter the solution. The solution will then filter along the gradient into collecting pools which will then be pumped to on-site plants for further processing. Only some of the uranium (commonly about 70%) is actually extracted. The uranium concentrations within the solution are very important for the efficient separation of pure uranium from the acid. As different heaps will yield different concentrations, the solution is pumped to a mixing plant that is carefully monitored. The properly balanced solution is then pumped into a processing plant where the uranium is separated from the sulfuric acid. Heap leaching is significantly cheaper than traditional milling processes. The low costs allow for lower grade ore to be economically feasible (given that it is the right type of ore body). Environmental law requires that the surrounding ground water is continually monitored for possible contamination. The mine will also have to have continued monitoring even after the shutdown of the mine. In the past mining companies would sometimes go bankrupt, leaving the responsibility of mine reclamation to the public. Recent additions to the mining law require that companies set aside the money for reclamation before the beginning of the project. The money will be held by the public to insure adherence to environmental standards if the company were to ever go bankrupt.

In-situ leaching: In-situ leaching (ISL), also known as solution mining, or in-situ recovery (ISR) in North America, involves leaving the ore where it is in the ground, and recovering the minerals from it by dissolving them and pumping the pregnant solution to the surface where the minerals can be recovered. Consequently, there is little surface disturbance and no tailings or waste rock generated. However, the ore body needs to be permeable to the liquids used, and located so that they do not contaminate ground water away from the orebody. Uranium ISL uses the native groundwater in the orebody which is fortified with a complexing agent and in most cases an oxidant. It is then pumped through the underground orebody to recover the minerals in it by leaching. Once the pregnant solution is returned to the surface, the uranium is recovered in much the same way as in any other uranium plant (mill). In Australian ISL mines (Beverley, Four Mile and Honeymoon Mine) the oxidant used is hydrogen peroxide and the complexing agent is sulfuric acid. Kazakh ISL mines generally do not employ an oxidant but use much higher acid concentrations in the circulating solutions. ISL mines in the USA use an alkali leach due to the presence of significant quantities of acid-consuming minerals such as gypsum and limestone in the host aquifers. Any more than a few percent carbonate minerals means that alkali leach must be used in preference to the more efficient acid leach. The Australian government has published a best practice guide for in situ leach mining of uranium, which is being revised to take account of international differences.

Seawater recovery: The uranium concentration in seawater is low, approximately 3. 3 parts per billion or 3. 3 micrograms per liter of seawater. But the quantity of this resource is gigantic and some scientists believe this resource is practically limitless with respect to world-wide demand. That is to say, if even a portion of the uranium in seawater could be used the entire world's nuclear power generation fuel could be provided over a long time period. Some anti-nuclear proponents claim this statistic is exaggerated. Although research and development for recovery of this low-concentration element by inorganic adsorbents such as titanium oxide compounds has occurred since the 1960s in the United Kingdom, France, Germany, and Japan, this research was halted due to low recovery efficiency. At the Takasaki Radiation Chemistry Research Establishment of the Japan Atomic Energy Research Institute (JAERI Takasaki Research Establishment), research and development have continued culminating in the production of adsorbent by irradiation of polymer fiber. Adsorbents have been synthesized that have a functional group (amidoxime group) that selectively adsorbs heavy metals, and the performance of such adsorbents has been improved. Uranium adsorption capacity of the polymer fiber adsorbent is high, approximately tenfold greater in comparison to the conventional titanium oxide adsorbent. In 2012, it was estimated that this fuel source could be extracted at 10 times the current price of uranium. In 2014, with the advances made in the efficiency of seawater uranium extraction, it was suggested that it would be economically competitive to produce fuel for light water reactors from seawater if the process was implemented at large scale. Uranium extracted on an industrial scale from seawater would constantly be replenished by both river erosion of rocks and the natural process of uranium dissolved from the surface area of the ocean floor, both of which maintain the solubility equilibria of seawater concentration at a stable level. Some commentators have argued that this strengthens the case for nuclear power to be considered a renewable energy.

## 9.3 Milling

The fundamental purpose of milling is to receive ore materials from the mining operation and extract uranium from it, yielding a higher uranium density and more useable form called "yellowcake". Uranium mills are typically located near its associated mine to reduce transportation costs, since there can be a large weight reduction during the milling process. The milling process is quite similar when preceded by open-pit or underground mining operations; however, the process is reduced if preceded by ISL since the end product of ISL yields a chemical solution rather than solid ore materials.

During milling the uranium bearing ore is initially crushed (if preceded by conventional mining), which is typically a dry process, and then ground into a slurry, which is a wet process. The intent of crushing and grinding is to effectively increase the surface area to volume ratio of the material, which facilitates chemical leaching. Two stages of crushing may be used: primary (yields ≈ 15 cm material) and fine crushing (yields ≈ 1 cm material). Crushing is typically accomplished by semi-autogenous grinding or autogenous grinding.

The material is typically roasted to remove a large portion of organic matter, such as sulfides and carbon-bearing compounds. Note that $UO_3$ is stable in air up to about 668 ℃, which then decomposes to $U_3O_8$ at higher temperatures. Next, the slurry is leached in solution, whereby the solution is selected depending on the ore materials. Sulfuric acid is the most commonly used solution, in addition to various carbonates. Much of this process is akin to ISL. One advantage of using ISL is that crushing is avoided, which reduces dust production and thereby reduces radon gas release, which is a radioactivity concern. The next stage in the process requires separation of liquids and solids, which is typically a relatively expensive process. A number of methods have been used to achieve this objective, including counter-current decants, drum filters, high-rate thickeners, cyclones, and horizontal belt filters. After solids have been separated from solution, the remaining liquid solution must be purified. The precise process that is taken depends on the composition of the solution, which depends principally on the original ore material and the desired purity of the end product. This is typically accomplished through ion-exchange or solvent extraction. The end product of milling is typically yellowcake, which principally contains $U_3O_8$ in addition to some impurities.

### 词汇

| | |
|---|---|
| copper | 铜 |
| ore | 矿石 |
| in-situ leach | 原位浸出 |
| underground mining | 地下开采 |
| open pit | 露天开采 |
| heap leaching | 堆浸 |
| co-product/by-product | 副产品 |
| groundwater | 地下水 |
| density | 密度 |
| fund | 基金 |
| reclamation | 改造 |
| surface | 表面 |

| | |
|---|---|
| tunnel | 地下通道 |
| uranium ore | 铀矿 |
| excavation | 挖掘 |
| rock | 岩石 |
| cement | 水泥 |
| radon exposure | 氡暴露 |
| hazard | 危险 |
| mine tailing | 尾矿砂 |
| volume | 体积 |
| chemicals | 化学品 |
| oxidation | 氧化 |
| process | 流程 |
| gradient | 坡度 |
| plastic | 塑料 |
| leaching agent | 浸取剂 |
| filter | 过滤器 |
| concentration | 浓集 |
| separation | 分离 |
| acid | 酸 |
| solution mining | 溶浸采矿 |
| in-site recovery(ISR) | 现场恢复 |
| dissolve | 溶解 |
| disturbance | 干扰 |
| oxidant | 氧化剂 |
| hydrogen peroxide | 过氧化氢 |
| complexing agent | 络合剂 |
| alkali leach | 碱浸 |
| microgram | 微克 |
| liter | 公升 |
| adsorbent | 吸附剂 |
| irradiation | 辐照 |
| polymer fiber | 聚合物纤维 |
| solubility equilibria | 溶解度平衡 |
| stable | 稳固的 |
| nuclear power | 核能 |

| | |
|---|---|
| renewable energy | 可再生能源 |
| milling | 提取 |
| yellowcake | 黄饼 |
| crush | 压碎 |
| slurry | 泥浆 |
| autogenous | 自生的 |
| grinding | 磨碎 |
| roast | 烤 |
| sulfide | 硫化物 |
| decompose | 分解 |
| dust | 灰尘 |
| decant | 注入 |
| thickener | 增稠剂 |
| cyclone | 气旋 |

**注释**

1. Uranium is found throughout the Earth's crust in various concentrations with an average abundance of approximately 2. 8 g/t.

铀以不同浓度存在于整个地壳中,其平均浓度约为 2. 8 克/吨。

2. The production cost of uranium depends on many factors, including the method used to mine the ore, geography, geology, the quality of grade of the ore, and the mineral type.

铀的生产成本取决于许多因素,包括开采矿石的方法、地理条件、地质条件、矿石的品位质量和矿物类型。

3. In 2016, the percentage of the mined uranium produced by each mining method was: in-situ leaching (49. 7%), underground mining (30. 8%), open pit (12. 9%), heap leaching (0. 4%), co-product/by-product (6. 1%). The remaining 0. 1% was derived as miscellaneous recovery.

2016 年,每种采矿方法开采的铀的百分比为:原位浸出(49. 7%)、地下开采(30. 8%)、露天开采(12. 9%)、堆浸(0. 4%)、副产品(6. 1%)。其余 0. 1%作为杂项回收。

4. Uranium extracted on an industrial scale from seawater would constantly be replenished by both river erosion of rocks and the natural process of uranium dissolved from the surface area of the ocean floor, both of which maintain the solubility equilibria of seawater concentration at a stable level.

由于河流对岩石的侵蚀和海底表面铀的自然溶解,以工业规模从海水中提取的铀可以不断得到补充,从而实现海水中铀的溶解度平衡保持在稳定水平。

5. The fundamental purpose of milling is to receive ore materials from the mining operation and extract uranium from it, yielding a higher uranium density and more useable form called "yellowcake".

铀的提取的基本目的是通过采矿作业接收矿石材料并从中提取铀,从而得到更高的铀密度和更有用的被称为“黄饼”的产物。

6. The material is typically roasted to remove a large portion of organic matter, such as sulfides and carbon-bearing compounds.

这种材料通常经过烘烤以去除大部分有机物,如硫化物和含碳化合物。

# Chapter 10 Uranium Conversion and Enrichment

## 10.1 Uranium Conversion

Following the mining and milling of uranium ore, an impure uranium ore concentrate (UOC), which is often referred to as "yellowcake" is produced. Depending on the ore type and the milling process used, the UOC chemical forms may be different and can include triuranium octaoxide ($U_3O_8$), uranyl peroxide ($UO_4$), or to a lesser degree ammonium, sodium, or magnesium diuranates. Uranium conversion, which also incorporates uranium refining, is a necessary processing step for the production of nuclear fuel. The primary purpose of refining is to remove impurities from the uranium that could lead to deleterious effects during fuel manufacturing, enrichment, and irradiation in a nuclear power reactor. The objective of conversion is to produce uranium in a chemical form suitable for enrichment (i. e., $UF_6$) or a form for direct fuel fabrication for heavy-water reactors (i. e., $UO_2$, since heavy water reactors do not require enrichment). Typically, the sequence of operations is refining first followed by conversion. However, at the Honeywell Metropolis Works Facility in the United States, the sequence is reversed, with conversion followed by refining. Since the majority of commercial nuclear power reactors in the world require enriched fuel, essentially all commercial fuel conversion facilities are dedicated to refining and production of $UF_6$. In general, there are two methods used for the production of $UF_6$— the "wet" conversion process and the "dry" fluoride volatility process.

**Wet process**

The main, "wet" process, is used by Cameco in Canada, by Orano in France, at Lanzhou in China and Seversk in Russia. For the wet process, the concentrate is first dissolved in nitric acid. The resulting clean solution of uranyl nitrate $UO_2(NO_3)_2 \cdot 6H_2O$ is fed into a countercurrent solvent extraction process, using tributyl phosphate dissolved in kerosene or dodecane. The uranium is collected by the organic extractant, from which it can be washed out by dilute nitric acid solution and then concentrated by evaporation. The solution is then calcined in a fluidized bed reactor to produce $UO_3$(or $UO_2$ if heated sufficiently). Alternatively, the uranyl nitrate may be concentrated and have ammonia injected to produce ammonium diuranate, which is then calcined to produce pure $UO_3$. Crushed $U_3O_8$ from the dry process and purified uranium oxide $UO_3$ from the wet process are then reduced in a kiln by hydrogen to $UO_2$:

$$U_3O_8 + 2H_2 \longrightarrow 3UO_2 + 2H_2O \quad \Delta H = -109 \text{ kJ/mol} \qquad (10.1)$$

or

$$UO_3 + H_2 \longrightarrow UO_2 + H_2O \quad \Delta H = -109 \text{ kJ/mol} \tag{10.2}$$

This reduced oxide is then reacted in another kiln with gaseous hydrogen fluoride (HF) to form uranium tetrafluoride ($UF_4$), though in some places this is made with aqueous HF by a wet process:

$$UO_2 + 4HF \longrightarrow UF_4 + 2H_2O \quad \Delta H = -176 \text{ kJ/mol} \tag{10.3}$$

The tetrafluoride is then fed into a fluidized bed reactor or flame tower with gaseous fluorine to produce uranium hexafluoride, $UF_6$. Hexafluoride ("hex") is condensed and stored.

$$UF_4 + F_2 \longrightarrow UF_6 \tag{10.4}$$

Removal of impurities takes place at each step.

### Dry process

The dry fluoride volatility route is unique to Honeywell in the United States. The process does have similarities to the wet conversion process, however with one major difference; uranium purification is at the end of the process rather than the beginning. In the dry process, uranium oxide concentrates are first calcined (heated strongly) to drive off some impurities, then agglomerated and crushed. At Converdyn's US conversion plant, $U_3O_8$ is first made into impure $UF_6$ and this is then refined in a two-stage distillation process.

$UF_6$, particularly if moist, is highly corrosive. When warm it is a gas, suitable for use in the enrichment process. At lower temperature and under moderate pressure, the $UF_6$ can be liquefied. The liquid is run into specially designed steel shipping cylinders which are thick walled and weigh over 15 tonnes when full. As it cools, the liquid $UF_6$ within the cylinder becomes a white crystalline solid and is shipped in this form (see Figure 10.1).

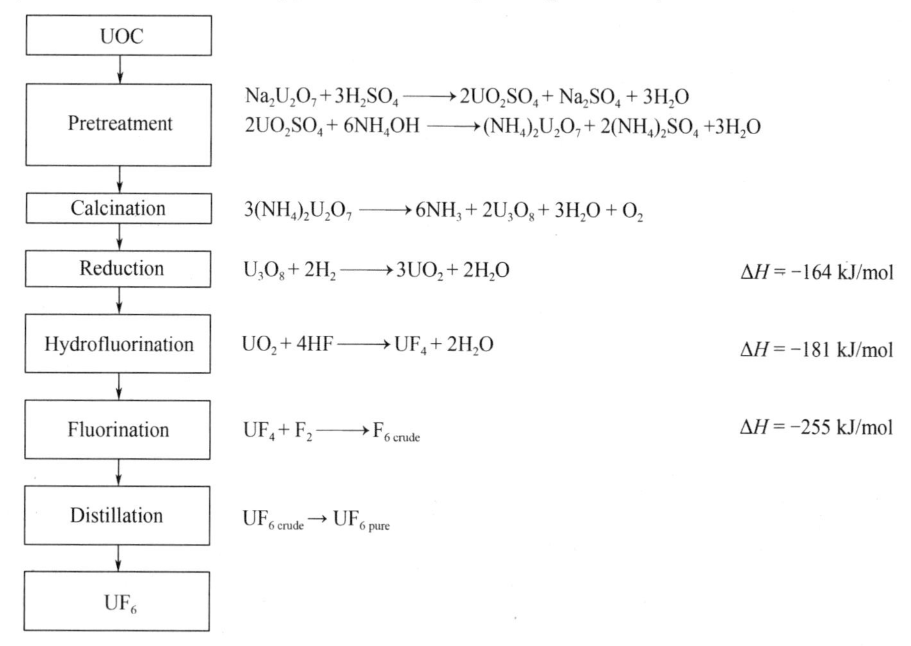


**Figure 10.1 Conversion of UOC to $UF_6$ using the "dry process"**

The siting, environmental and security management of a conversion plant is subject to the regulations that are in effect for any chemical processing plant involving fluorine-based chemicals.

## 10.2 Uranium Enrichment

Uranium hexafluoride ($UF_6$) is the feed material used in all commercial enrichment processes around the world. It is an ideal chemical for enrichment processes due to its physicochemical properties. It has a convenient triple point at 64 ℃ and 152 kPa, where manipulation of $UF_6$ as a gas, liquid, or solid is achieved by subtle manipulations of temperature and pressure. A second advantage of $UF_6$ is that the fluorine atom only has one naturally occurring isotope, which makes uranium isotope separation processes based on mass difference feasible. Since all fluorine atoms in $UF_6$ are the same, any mass difference in $UF_6$ must be due to different uranium isotopes.

Uranium enrichment technologies are well-known, and many have been developed since the Manhattan Project in the 1940s. The two commercial-scale enrichment processes that have been successfully demonstrated are gaseous diffusion and gas centrifuge enrichment. Since 2013, enrichment by gaseous diffusion has been discontinued and fully replaced by the gas centrifuge process. The retirement of gaseous diffusion plants was largely due to their energy intensive nature. Currently, all commercial enrichment facilities utilize a gas centrifuge process, which require less electricity than gas diffusion plants and are generally far less expensive. However, developments in laser enrichment technologies suggest that it may be the next-generation enrichment technology. A key feature for consideration of laser enrichment is its low energy intensity relative to that of the gas centrifuge.

### Gas centrifuge enrichment process

Gas centrifuge enrichment technology was originally developed in the early 1940s, at the same time as other technologies were being developed. However, it was not advanced at the time due to technical disadvantages and the availability of the simpler gaseous diffusion process. After World War Ⅱ, additional development and significant technical advances were made to this technology, such that gas centrifuge technology was reintroduced during the 1960s and took its place as the second-generation enrichment technology. Since that time, the implementation of this technology has grown to the point that it is now the sole technology used at commercial enrichment facilities.

A gas centrifuge is a long, narrow, vertical cylinder known as a rotor, which is rapidly rotated on its axis within an evacuated casing. A schematic of a centrifuge, based on the Zippe design from the early 1960s, is illustrated in Figure 10.2. The operating principle of the gas centrifuge is centrifugal force operating at thousands of times that of gravity, which creates a density gradient in a gas mixture. In uranium enrichment, $UF_6$ gas is fed to the rotor, and the heavier $^{238}UF_6$ molecules move toward the outer wall, while the lighter $^{235}UF_6$ molecules gather at the centre of the cylinder. This partial separation is enhanced by an induced, axial, countercurrent flow of $UF_6$, which causes enriched and depleted streams to migrate toward opposite ends of the cylinder. The enriched and depleted streams are withdrawn from the cylinder via stationary fixtures at each end.

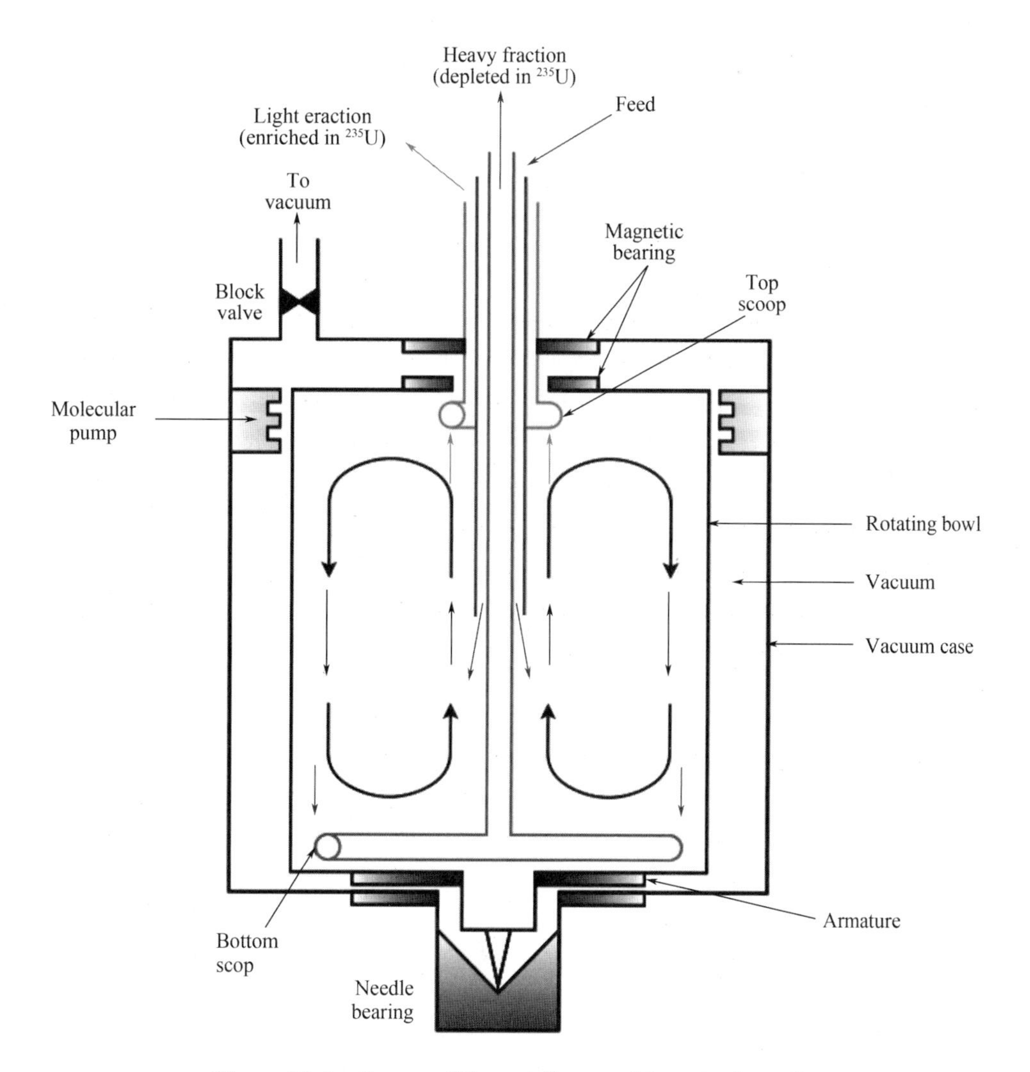


**Figure 10. 2 Gaseous $UF_6$ centrifuge enrichment schematic**

The separative capacity of a gas centrifuge improves with increases in two key design features: the length and rotational speed. At rotational speeds of up to 700 m/s, proper selection of materials of construction for the rotor is essential in order for it to withstand the mechanical stresses and the corrosive nature of $UF_6$. Increasing the length of the rotor requires precision fabrication capabilities, since the rotor must be finely balanced, straight, and of uniform wall thickness to prevent vibration and wobbling. Other factors that influence the separation factor include the axial location of the $UF_6$ feed point within the centrifuge, feed rate, and temperature.

The separation factor of a gas centrifuge is in the range of 1. 3−1. 6, and while mall, it is greater than separation factors for gaseous diffusion or aerodynamic processes. The process has to be repeated many times to achieve the desired $^{235}U$ concentration that is required for nuclear power plants. Centrifuges are connected in series, and the concentration of $^{235}U$ in the enriched stream is incrementally increased until the desired enrichment level is attained. Similarly, the depleted stream is further successively "stripped" in a series of centrifuges to reach a final $^{235}U$ concentration. In a cascade arrangement, centrifuges are also connected in parallel to attain the designed material flow rate for the enrichment plant.

### Gaseous diffusion process

The first commercial, large-scale process was the gaseous diffusion process. Developed in the early 1940s during World War Ⅱ, it became the first-generation enrichment technology. The general principle behind gaseous diffusion exploits the mass difference between the two uranium isotopes. At thermal equilibrium, the average kinetic energies of gases in a mixture are the same. Since the kinetic energy of a molecule is a function of its mass and velocity, molecules with the same average kinetic energy have different average velocities that are related to their mass difference. On average, lighter molecules travel faster than heavier molecules and thus contact the walls of a vessel more frequently. If one of the walls is a membrane with holes large enough to allow single molecules to pass, but small enough to prevent bulk flow of the gas, more of the lighter molecules will pass through the holes than the heavier molecules.

Gaseous $UF_6$ diffusion enrichment separator is given in Figure 10. 3.

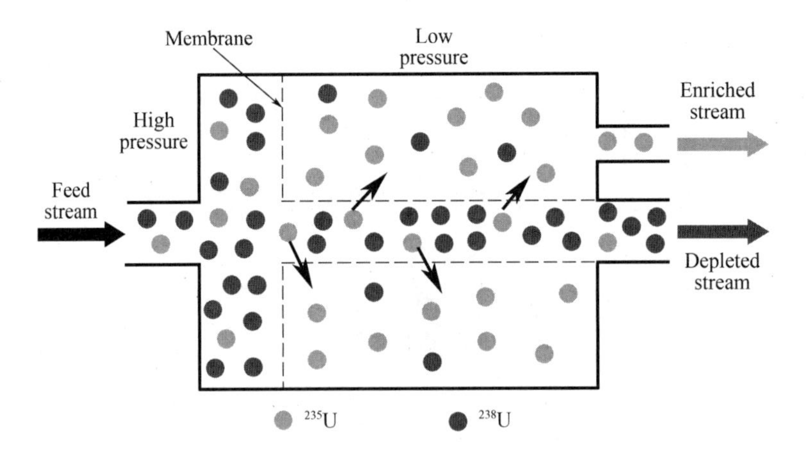


**Figure 10. 3 Gaseous $UF_6$ diffusion enrichment separator**

At an enrichment facility, each separation element consists of a converter, which is also known as a diffuser. The converter contains thousands of tubes of diffusion barriers. The diffusion barrier or membrane is perforated with submicron holes. Gaseous $UF_6$ is pumped to higher pressure by primary and secondary compressors and then fed to a cooler where the heat created from gas compression is removed. The cooled gas then flows to the inside of the barrier tubes, and the faster moving $^{235}UF_6$ molecules more readily pass through the holes into the lower pressure area of the converter. Thus the gas diffusing through these tubes is partially enriched, having a slightly higher concentration of $^{235}U$. The diffused gas is then fed to the next stage, and the process is repeated. The remaining, partially depleted, stream from the converter is fed to another stage where the material is treated further.

The separation factor of a gaseous diffusion stage for $UF_6$ is very small and in the range of 1.0040-1.0045 due to the very small mass difference between the two uranium isotopes. Thus, gaseous diffusion cascades consist of many separation stages.

**词汇**

| | |
|---|---|
| uranium ore concentrate | 铀矿浓缩 |
| triuranium octaoxide | 八氧化三铀 |
| uranyl peroxide | 过氧化铀 |
| diuranate | 重铀酸盐 |
| uranium conversion | 铀的转化 |
| fuel manufacturing | 燃料制造 |
| enrichment | 浓缩 |
| irradiation | 放射 |
| chemical form | 化学形态 |
| fuel fabrication | 燃料制造 |
| heavy-water reactor | 重水堆 |
| uranyl nitrate | 硝酸铀酰 |
| tributyl phosphate | 磷酸三丁酯 |
| kerosene | 煤油 |
| dodecane | 十二烷 |
| organic extractant | 有机萃取剂 |
| dilute nitric acid | 稀硝酸 |
| evaporation | 蒸发 |
| calcined | 煅烧 |
| hydrogen | 氢气 |
| fluoride volatility | 氟化挥发法 |
| distillation | 蒸馏 |
| corrosive | 腐蚀性的 |
| uranium hexafluoride | 六氟化铀 |
| uranium isotope | 铀同位素 |
| gaseous diffusion | 气体扩散 |
| gas centrifuge process | 气体离心法 |
| gas centrifuge enrichment technology | 气体离心浓缩技术 |
| separative capacity | 分离能力 |

| | |
|---|---|
| flow rate | 流率;流量 |
| gaseous diffusion process | 气体扩散法 |
| diffusion barrier | 扩散膜 |
| separation factor | 分离系数 |

**注释**

1. Uranium conversion, which also incorporates uranium refining, is a necessary processing step for the production of nuclear fuel. The primary purpose of refining is to remove impurities from the uranium that could lead to deleterious effects during fuel manufacturing, enrichment, and irradiation in a nuclear power reactor.

铀纯化是铀转化的一部分,是生产核燃料的必要加工步骤。纯化的主要目的是去除铀中的杂质,这些杂质可能会在燃料制造、浓缩和辐照过程中对核反应堆产生有害影响。

2. The objective of conversion is to produce uranium in a chemical form suitable for enrichment (i. e., $UF_6$) or a form for direct fuel fabrication for heavy-water reactors (i. e., $UO_2$, since heavy water reactors do not require enrichment).

铀转化的目的是以 $UF_6$ 或 $UO_2$ 的形式来生产铀。$UF_6$ 便于浓缩;$UO_2$ 可在重水堆中直接使用。

3. The two commercial-scale enrichment processes that have been successfully demonstrated are gaseous diffusion and gas centrifuge enrichment.

人们已经成功证实,可实现商业规模生产的两种浓缩工艺是气体扩散法和气体离心法。

4. However, developments in laser enrichment technologies suggest that it may be the next-generation enrichment technology. A key feature for consideration of laser enrichment is its low energy intensity relative to that of the gas centrifuge.

然而,激光浓缩技术的发展表明,它可能是下一代浓缩技术。激光浓缩技术的一个关键特征是,相对于气体离心机而言,其对能源的需求更少。

# Chapter 11 Nuclear Fuel

Nuclear fuel is material used in nuclear power stations to produce heat to power turbines. Heat is created when nuclear fuel undergoes nuclear fission. Most nuclear fuels contain heavy fissile actinide elements that are capable of undergoing and sustaining nuclear fission. The three most relevant fissile isotopes are $^{233}U$, $^{235}U$ and $^{239}Pu$. When the unstable nuclei of these atoms are hit by a slow-moving neutron, they frequently split, creating two daughter nuclei and two or three more neutrons. In that case, the neutrons released go on to split more nuclei. This creates a self-sustaining chain reaction that is controlled in a nuclear reactor, or uncontrolled in a nuclear weapon. Alternatively, if the nucleus absorbs the neutron without splitting, it creates a heavier nucleus with one additional neutron. The processes involved in mining, refining, purifying, using, and disposing of nuclear fuel are collectively known as the nuclear fuel cycle. Not all types of nuclear fuels create power from nuclear fission; $^{238}Pu$ and some other elements are used to produce small amounts of nuclear power by radioactive decay in radioisotope thermoelectric generators and other types of atomic batteries. Nuclear fuel has the highest energy density of all practical fuel sources.

## 11.1 Oxide Fuel

For fission reactors, the fuel (typically based on uranium) is usually based on the metal oxide; the oxides are used rather than the metals themselves because the oxide melting point is much higher than that of the metal and because it cannot burn, being already in the oxidized state.

Uranium dioxide: Uranium dioxide is a black semiconducting solid. It can be made by heating uranyl nitrate to form $UO_3$. This is then converted by heating with hydrogen to form $UO_2$. It can be made from enriched uranium hexafluoride by reacting with ammonia to form a solid called ammonium diuranate. This is then heated (calcined) to form $UO_3$ and $U_3O_8$ which is then converted by heating with hydrogen or ammonia to form $UO_2$. The $UO_2$ is mixed with an organic binder and pressed into pellets, these pellets are then fired at a much higher temperature (in $H_2$/Ar) to sinter the solid. The aim is to form a dense solid which has few pores. The thermal conductivity of uranium dioxide is very low compared with that of zirconium metal, and it goes down as the temperature goes up. Corrosion of uranium dioxide in water is controlled by similar electrochemical processes to the galvanic corrosion of a metal surface. While exposed to the neutron flux during normal operation in the core environment a small percentage of the $^{238}U$ in the fuel absorbs excess neutrons and is transmuted into $^{239}U$. $^{239}U$ rapidly decays into $^{239}Np$ which in turn rapidly decays into $^{239}Pu$. The small percentage of $^{239}Pu$ has a higher neutron cross section than $^{235}U$. As the $^{239}Pu$ accumulates the chain reaction shifts from pure $^{235}U$ at initiation of the fuel use to a ratio of about 70% $^{235}U$ and 30% $^{239}Pu$ at the end of the 18 to 24 month fuel exposure period.

MOX: Mixed oxide, or MOX fuel, is a blend of plutonium and natural or depleted uranium which behaves similarly (though not identically) to the enriched uranium feed for which most nuclear reactors were designed. MOX fuel is an alternative to low enriched uranium (LEU) fuel used in the light water reactors which predominate nuclear power generation.

Some concern has been expressed that used MOX cores will introduce new disposal challenges, though MOX itself is a means to dispose of surplus plutonium by transmutation.

Reprocessing of commercial nuclear fuel to make MOX was done in the Sellafield MOX Plant (England). As of 2015, MOX fuel is made in France (see Marcoule Nuclear Site), and to a lesser extent in Russia, India and Japan. China plans to develop fast breeder reactors and reprocessing.

The Global Nuclear Energy Partnership, was a U. S. proposal in the George W. Bush Administration to form an international partnership to see spent nuclear fuel reprocessed in a way that renders the plutonium in it usable for nuclear fuel but not for nuclear weapons. Reprocessing of spent commercial-reactor nuclear fuel has not been permitted in the United States due to nonproliferation considerations. All of the other reprocessing nations have long had nuclear weapons from military-focused "research"-reactor fuels except for Japan. Normally, with the fuel being changed every three years or so, about half of the $^{239}Pu$ is "burned" in the reactor, providing about one third of the total energy. It behaves like $^{235}U$ and its fission releases a similar amount of energy. The higher the burn-up, the more plutonium in the spent fuel, but the lower the fraction of fissile plutonium. Typically about 1% of the used fuel discharged from a reactor is plutonium, and some two thirds of this is fissile (about 50% $^{239}Pu$, 15% $^{241}Pu$). Worldwide, some 70 tonnes of plutonium contained in used fuel is removed when refueling reactors each year.

## 11.2 Metal Fuel

Metal fuels have the advantage of a much higher heat conductivity than oxide fuels, but cannot survive equally high temperatures. Metal fuels have a long history of use, stretching from the Clementine reactor in 1946 to many test and research reactors. Metal fuels have the potential for the highest fissile atom density. Metal fuels are normally alloyed, but some metal fuels have been made with pure uranium metal. Uranium alloys that have been used include uranium aluminum, uranium zirconium, uranium silicon, uranium molybdenum, and uranium zirconium hydride (UZrH). Any of the aforementioned fuels can be made with plutonium and other actinides as part of a closed nuclear fuel cycle. Metal fuels have been used in water reactors and liquid metal fast breeder reactors, such as EBR-Ⅱ.

TRIGA fuel: TRIGA fuel is used in TRIGA (Training, Research, Isotopes, General Atomics) reactors. The TRIGA reactor uses UZrH fuel, which has a prompt negative fuel temperature coefficient of reactivity, meaning that as the temperature of the core increases, the reactivity decreases, so it is highly unlikely for a meltdown to occur. Most cores that use this fuel are "high leakage" cores where the excess leaked neutrons can be utilized for research. That is, they can be used as a neutron source. TRIGA fuel was originally designed to use highly enriched uranium, however in 1978 the U. S. Department of Energy launched its Reduced Enrichment for Research Test Reactors program, which promoted reactor conversion to low-enriched uranium fuel. A total of 35 TRIGA reactors have been installed at locations across the U. S. A further 35 reactors have been installed in other countries.

Actinide fuel: In a fast neutron reactor, the minor actinides produced by neutron capture of uranium and plutonium can be used as fuel. Metal actinide fuel is typically an alloy of zirconium, uranium, plutonium, and minor actinides. It can be made inherently safe as thermal expansion of the metal alloy will increase neutron leakage.

Molten plutonium: Molten plutonium, alloyed with other metals to lower its melting point and encapsulated in tantalum, was tested in two experimental reactors, LAMPRE Ⅰ and LAMPRE Ⅱ, at Los Alamos National Laboratory in the 1960s. "LAMPRE experienced three separate fuel failures during operation."

## 11.3 Non-oxide Ceramic Fuels

Ceramic fuels other than oxides have the advantage of high heat conductivities and melting points, but they are more prone to swelling than oxide fuels and are not understood as well.

Uranium nitride: This is often the fuel of choice for reactor designs that NASA produces, one advantage is that UN has a better thermal conductivity than $UO_2$. Uranium nitride has a very high melting point. This fuel has the disadvantage that unless $^{15}N$ was used (in place of the more common $^{14}N$) that a large amount of $^{14}C$ would be generated from the nitrogen by the (n, p) reaction. As the nitrogen required for such a fuel would be so expensive it is likely that the fuel would have to be reprocessed by pyroprocessing to enable the $^{15}N$ to be recovered. It is likely that if the fuel was processed and dissolved in nitric acid that the nitrogen enriched with $^{15}N$ would be diluted with the common $^{14}N$. Fluoride volatility is a method of reprocessing that does not rely on nitric acid, but it has only been demonstrated in relatively small scale installations whereas the established PUREX process is used commercially for about a third of all spent nuclear fuel (the rest being largely subject to a "once through fuel cycle"). All nitrogen-fluoride compounds are volatile or gaseous at room temperature and could be fractionally distilled from the other gaseous products (including recovered uranium hexafluoride) to recover the initially used nitrogen. If the fuel could be processed in such a way as to ensure low contamination with non-radioactive carbon (not a common fission product and absent in nuclear reactors that don't use it as a moderator) then Fluoride volatility could be used to separate the $^{14}C$ produced by producing carbon tetrafluoride. $^{14}C$ is proposed for use in particularly long lived low power nuclear batteries called diamond batteries.

Uranium carbide: Much of what is known about uranium carbide is in the form of pin-type fuel elements for liquid metal fast reactors during their intense study during the 1960s and 1970s. However, recently there has been a revived interest in uranium carbide in the form of plate fuel and most notably, micro fuel particles (such as TRISO particles). The high thermal conductivity and high melting point makes uranium carbide an attractive fuel. In addition, because of the absence of oxygen in this fuel (during the course of irradiation, excess gas pressure can build from the formation of $O_2$ or other gases) as well as the ability to complement a ceramic coating (a ceramic-ceramic interface has structural and chemical advantages), uranium carbide could be the ideal fuel candidate for certain Generation Ⅳ reactors such as the gas-cooled fast reactor. While the neutron cross section of carbon is low, during years of burnup, the predominantly $^{12}C$ will undergo neutron capture to produce stable $^{13}C$ as well as radioactive $^{14}C$. Unlike the $^{14}C$ produced by using Uranium nitrate, the $^{14}C$ will make up only a small isotopic impurity in the overall carbon content and thus make the entirety of the carbon content unsuitable for non-nuclear uses but the $^{14}C$ concentration will be too low for use in nuclear batteries without enrichment. Nuclear graphite discharged from reactors where it was used as a moderator presents the same issue.

## 11.4 Liquid Fuels

Liquid fuels are liquids containing dissolved nuclear fuel and have been shown to offer numerous operational advantages compared to traditional solid fuel approaches. Liquid-fuel reactors offer significant safety advantages due to their inherently stable "self-adjusting" reactor dynamics. This provides two major benefits: virtually eliminating the possibility of a run-away reactor meltdown, providing an automatic load-following capability which is well suited to electricity generation and high-temperature industrial heat applications.

Another major advantage of the liquid core is its ability to be drained rapidly into a passively safe dump-tank. This advantage was conclusively demonstrated repeatedly as part of a weekly shutdown procedure during the highly successful 4 a molten salt reactor (MSR) Experiment. Another huge advantage of the liquid core is its ability to release xenon gas which normally acts as a neutron absorber ($^{135}Xe$ is the strongest known neutron poison and is produced both directly and as a decay product of $^{135}I$ as a fission product) and causes structural occlusions in solid fuel elements (leading to the early replacement of solid fuel rods with over 98% of the nuclear fuel unburned, including many long-lived actinides). In contrast, molten salt reactors are capable of retaining the fuel mixture for significantly extended periods, which not only increases fuel efficiency dramatically but also incinerates the vast majority of its own waste as part of the normal operational characteristics. A downside to letting the $^{135}Xe$ escape instead of allowing it to capture neutrons converting it to the basically stable and chemically inert $^{136}Xe$, is that it will quickly decay to the highly chemically reactive long-lived radioactive $^{135}Cs$, which behaves similar to other alkali metals and can be taken up by organisms in their metabolism.

Molten salt fuel: Molten salt fuels have nuclear fuel dissolved directly in the molten salt coolant. Molten salt-fueled reactors, such as the liquid fluoride thorium reactor (LFTR), are different from molten salt-cooled reactors that do not dissolve nuclear fuel in the coolant. Molten salt fuels were used in the LFTR known as the Molten Salt Reactor Experiment, as well as other liquid core reactor experiments. The liquid fuel for the molten salt reactor was a mixture of lithium, beryllium, thorium and uranium fluorides: $LiF-BeF_2-ThF_4-UF_4$(72-16-12-0.4 mol%). It had a peak operating temperature of 705 ℃ in the experiment, but could have operated at much higher temperatures since the boiling point of the molten salt was in excess of 1 400 ℃.

Aqueous solutions of uranyl salts: The aqueous homogeneous reactors (AHRs) use a solution of uranyl sulfate or other uranium salt in water. Historically, AHRs have all been small research reactors, not large power reactors. An AHR known as the Medical Isotope Production System is being considered for production of medical isotopes.

Liquid metals or alloys: The Dual fluid reactor has a variant DFR/m which works with eutectic liquid metal alloys, e.g., U-Cr or U-Fe.

## 11.5 Accident Tolerant Fuels

Accident tolerant fuels (ATF) are a series of new nuclear fuel concepts, researched in order to improve fuel performance under accident conditions, such as loss-of-coolant accident (LOCA) or reaction-initiated accidents (RIA). These concerns became more prominent after the Fukushima Daiichi nuclear disaster in Japan, in particular regarding light-water reactor (LWR) fuels performance under accident conditions. The aim of the research is to develop nuclear fuels that can tolerate loss of active cooling for a considerably longer period than the existing fuel designs and prevent or delay the release of radionuclides during an accident. This research is focused on reconsidering the design of fuel pellets and cladding, as well as the interactions between the two.

### 词汇

| | |
|---|---|
| oxide | 氧化物 |
| melting point | 熔点 |
| oxidized state | 氧化态 |
| uranium dioxide | 二氧化铀 |
| solid | 固体 |
| uranyl nitrate | 硝酸铀酰 |
| hydrogen | 氢气 |
| enriched uranium hexafluoride | 浓缩六氟化铀 |
| ammonium diuranate | 重铀酸铵 |
| thermal conductivity | 导热性 |
| corrosion | 腐蚀 |
| electrochemical process | 电化学过程 |
| neutron flux | 中子通量 |
| neutron cross section | 中子截面 |
| Mixed oxide, or MOX fuel | 混合氧化物,或混合氧化物燃料 |
| fast breeder reactor | 快中子增殖反应堆 |
| nonproliferation | 防扩散 |

| | |
|---|---|
| ton | 吨 |
| metal fuel | 金属燃料 |
| heat conductivity | 热导率 |
| uranium alloy | 铀合金 |
| reactivity | 反应性 |
| actinide fuel | 锕系燃料 |
| fast neutron reactor | 快中子反应堆 |
| neutron capture | 中子俘获 |
| thermal expansion | 热膨胀 |
| ceramic fuel | 陶瓷燃料 |
| swelling | 肿胀 |
| uranium nitride | 氮化铀 |
| pyroprocessing | 干法后处理 |
| nuclear battery | 核电池 |
| uranium carbide | 碳化铀 |
| oxygen | 氧气 |
| irradiation | 照射 |
| generation Ⅳ reactors | 第四代反应堆 |
| gas-cooled fast reactor | 气冷快堆 |
| reactor meltdown | 反应堆熔毁 |
| electricity generation | 发电 |
| neutron poison | 中子毒物 |
| fission product | 裂变产物 |
| Molten Salt Reactors (MSR) | 熔盐堆 |
| inert | 惰性的 |
| alkali metal | 碱金属 |
| organism | 有机体 |
| metabolism | 新陈代谢 |
| molten salt fuel | 熔盐燃料 |
| accident tolerant fuels (ATF) | 事故容错燃料 |
| fuel performance | 燃料性能 |
| loss-of-coolant accident(LOCA) | 冷却剂损失事故 |

**注释**

1. For fission reactors, the fuel (typically based on uranium) is usually based on the metal oxide; the oxides are used rather than the metals themselves because the oxide melting point is much higher than that of the metal and because it cannot burn, being already in the oxidized state.

对于裂变反应堆,燃料(通常以铀为基础)通常以金属氧化物为主;使用金属氧化物而不使用金属本身,是因为氧化物的熔点比金属的熔点高得多,并且由于它已经处于氧化状态,因此不会燃烧。

2. Uranium dioxide is a black semiconducting solid. It can be made by heating uranyl nitrate to form $UO_3$. This is then converted by heating with hydrogen to form $UO_2$.

二氧化铀是一种黑色的固态半导体。它可以通过加热硝酸铀酰形成 $UO_3$,然后通过在氢气作用下加热将其转化为 $UO_2$。

3. Metal fuels have the advantage of a much higher heat conductivity than oxide fuels but cannot survive equally high temperatures.

金属燃料具有比氧化物燃料高得多的热导率,但金属燃料不能像氧化物燃料那样在较高的温度下存在。

4. In a fast neutron reactor, the minor actinides produced by neutron capture of uranium and plutonium can be used as fuel.

在快中子反应堆中,铀和钚的中子捕获所产生的少量锕系元素可以用作燃料。

5. Ceramic fuels other than oxides have the advantage of high heat conductivities and melting points, but they are more prone to swelling than oxide fuels and are not understood as well.

除氧化物以外的陶瓷燃料具有高导热性和高熔点的优点,但它们比氧化物燃料更容易膨胀,而人们对其机理并不了解。

6. As the nitrogen required for such a fuel would be so expensive it is likely that the fuel would have to be reprocessed by pyroprocessing to enable the $^{15}N$ to be recovered.

由于这种燃料所需的氮气非常昂贵,因此该燃料很可能必须通过高温干法工艺来对其进行再处理,以回收$^{15}N$。

7. Liquid fuels are liquids containing dissolved nuclear fuel and have been shown to offer numerous operational advantages compared to traditional solid fuel approaches.

液态燃料是含有溶解核燃料的液体,与传统的固体燃料方法相比,液体燃料已被证明具有许多操作优势。

8. Liquid-fuel reactors offer significant safety advantages due to their inherently stable "self-adjusting" reactor dynamics.

液态燃料反应堆具有显著的安全优势,因为其固有的"自调节"反应堆的稳定性。

9. Accident tolerant fuels (ATF) are a series of new nuclear fuel concepts, researched in order to improve fuel performance under accident conditions, such as loss-of-coolant accident (LOCA) or reaction-initiated accidents (RIA).

事故容错燃料(ATF)是一系列新的核燃料概念,旨在提高事故条件下的燃料性能,如冷却剂损失事故(LOCA)或反应引发事故(RIA)。

10. The aim of the research is to develop nuclear fuels that can tolerate loss of active cooling for a considerably longer period than the existing fuel designs and prevent or delay the release of radionuclides during an accident.

这项研究的目的是开发一些核燃料,使其相比于现有燃料,能够在更长时间里承受主动冷却损失,并在事故期间防止或延迟放射性核素的释放。

# Chapter 12 Nuclear Power Reactors

## 12.1 Nuclear Reactor Types

Nuclear reactors can be divided into several types with respect to the use of nuclear fuel and the manufacture of fuel elements. The various types of nuclear reactors are distinguished on the basis of the following aspects:

(1) the kind of fuel used (e. g., natural U, enriched U, Pu, U/Pu mixtures);

(2) the energy of the neutrons used for fission (thermal or fast reactors);

(3) the kind of moderator (e. g., graphite, light water, heavy water);

(4) the combination of fuel and moderator (homogeneous or heterogeneous);

(5) the kind of coolant (e. g., gas, water, sodium, organic compounds, molten salts);

(6) the operation of the coolant (boiling water, pressurized water); and

(7) the application (e. g., research reactors, test reactors, power reactors, breeder reactors, converters, plutonium production, ship propulsion).

The first nuclear reactor was built by Fermi and co-workers, beneath the stand of a football stadium in Chicago, by using natural uranium and bars of graphite, and reached criticality in December 1942. It looked like a pile, the thermal power was 2 W, and cooling and radiation protection were not provided. The next nuclear reactor began operation in 1943 at Oak Ridge (USA) by using 54 t of uranium metal in the form of fuel rods inserted into a block of graphite 5.6 m long, shielded by concrete. Several reactors of a similar type (graphite-moderated, water-cooled, natural uranium reactors) were built in 1943 and in the following years at Hanford (USA) for the production of plutonium to be used as a nuclear explosive.

The concept of energy production by nuclear reactors has found greater interest since about 1950. The first nuclear power station [graphite-moderated, gas-cooled ($CO_2$), natural uranium] began operation in 1956 at Calder Hall (UK). Today, pressurized-water reactors and boiling-water reactors are the most widely used power reactors. They contain about 100 t of weakly enriched U (about 3.0% to 3.5% $^{235}U$). Fast breeder reactors contain about 100 t of natural U and about 6 t of Pu, but no moderator. They exhibit several advantages: the high burn-up is due to the fact that large amounts of $^{238}U$ are transformed into the easily fissile $^{239}Pu$. In this way, the energy production from U is increased by a factor of about 100, and enrichment of $^{235}U$ by isotope separation is not needed. Some problems are caused by use of liquid sodium as coolant due to its high reactivity. High-temperature gas-cooled reactors (HTGRs) have also some advantages, because the high temperature of the coolant gives high efficiency, but they have not found broad application. The operation of boiling-water reactors (BWRs), pressurized-water reactors (PWRs), and HTGRs are shown schematically in Figures 12.1 – 12.3, respectively. Power reactors have also been developed and installed for ship propulsion, for instance, in submarines (e. g. , Nautilus, USA) or icebreakers (e. g. , Lenin, Russia).

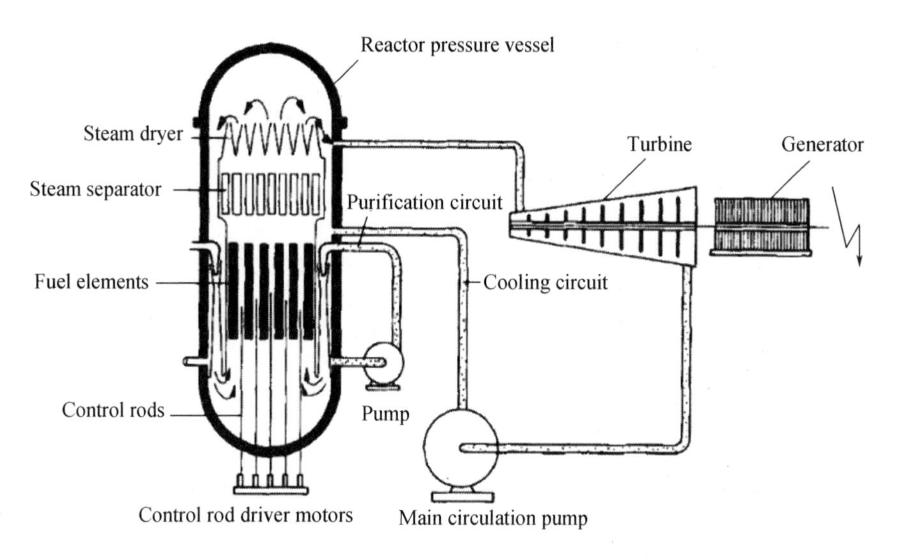


**Figure 12. 1　Boiling-water reactor (BWR)**

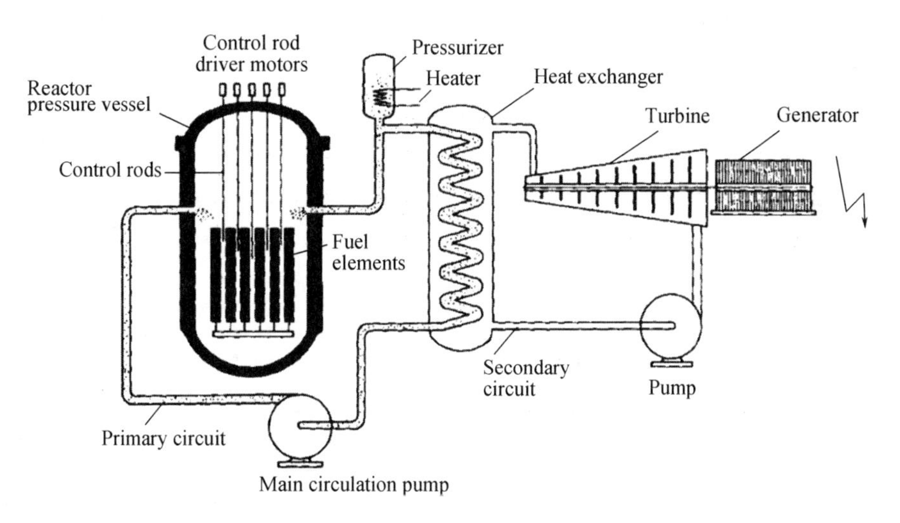


**Figure 12. 2　Pressurized-water reactor (PWR)**

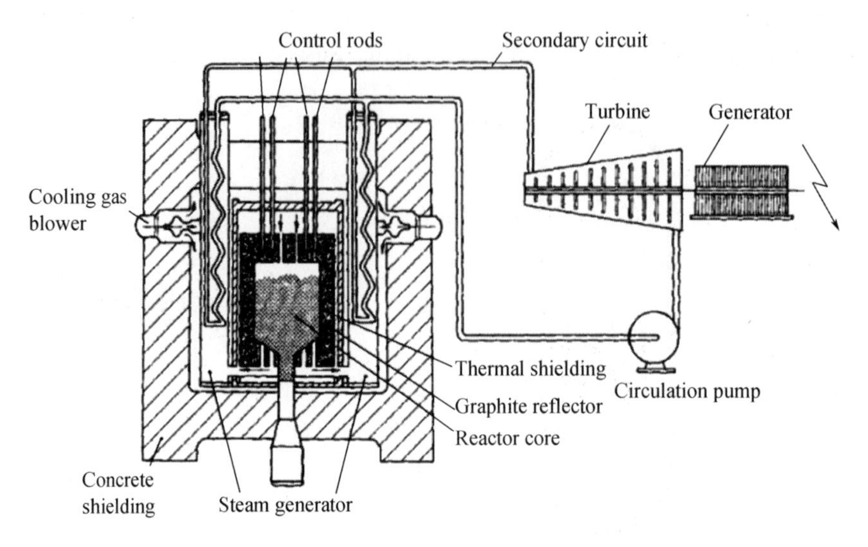


**Figure 12.3 High-temperature gas-cooled reactor (HTGR)**

## 12.2 Moderators and Coolants

The properties of some moderators and coolants are listed in Table 12.1. The purpose of the moderator is to take away the energy of the fission neutrons by collisions, without absorbing appreciable amounts of the neutrons. The absorption cross-section is relatively low for graphite, $D_2O$, $CO_2$, and He. The relatively high absorption cross-sections of $H_2O$ and liquid metals require use of enriched U or of Pu as fuel. Thermal conductivity is of special importance in power reactors. Application of the same materials as coolant and moderator is desirable in the case of thermal reactors.

**Table 12.1 Properties of some moderators and coolants**

| | Absorption cross-section for thermal neutrons $\sigma_a$ [barn] | Density at 20 °C [g · cm$^{-3}$] | Melting point [℃] | Boiling point [℃] | Thermal conductivity at 20 ℃ [J · cm$^{-1}$ · s$^{-1}$ · K$^{-1}$] | Specific heat at 20 ℃ [J · g$^{-1}$ · K$^{-1}$] |
|---|---|---|---|---|---|---|
| Graphite | 0.004 5 | 2.256 | Sublimation | 3 650 | 1.674 | 0.720 (25 ℃) |
| $D_2O$ | 0.001 1 | 1.105 | 3.8 | 101.42 | 0.005 86 | 4.212 |
| $H_2O$ | 0.66 | 0.998 | 0 | 100.0 | 0.005 86 | 4.183 |
| $CO_2$ | 0.003 8 | $1.977\times10^{-3}$ | Sublimation | −78.5 | 0.000 184 (30 ℃) | 0.833 (15 ℃) |
| He | 0.007 | $0.177\times10^{-3}$ | −272.2 | −268.6 | 0.000 611 (50 ℃) | 5.200 |
| Na | 0.53 | 0.928 (100 ℃) | 97.7 | 883 | 0.863 (100 ℃) | 1.386 (100 ℃) |

The disadvantages of water as a coolant are the low boiling temperature and the influence of corrosion. For operation at high temperatures, gases are preferable as coolants. As the ratio of heat transfer to pumping power is proportional to $M^2c_p^2$ ($M$ = molecular mass, $c_p$ = specific heat), hydrogen would be the most favorable coolant at high temperatures. However, because of its reactivity, use of hydrogen is prohibitive. Helium is rather expensive; $CO_2$ is suitable as a coolant for graphite-moderated reactors, but at high temperatures the instability of graphite due to the equilibrium $C(s) + CO_2(g) \rightleftharpoons 2\,CO(g)$ has to be taken into account. Liquid metals exhibit high thermal conductivity, but because of their reactivity, special precautions are necessary.

All other materials used in nuclear reactors for construction or as tubes should exhibit low neutron absorption, low activation, no change in properties under the influence of the high neutron and γ-ray fluxes, and high corrosion resistance. These requirements are best met by zirconium, which has found wide application in nuclear reactors. Al, Be, and Mg have limited applicability. Steel and other heavy metals are only applicable if their relatively high neutron absorption is acceptable.

The range of fission products is small (about 5−10 μm in solids and about 25 μm in water), but their specific ionization is high. This leads to high temperatures in solid fuel, in particular in $UO_2$ (up to several thousand degrees Celsius). Furthermore, lattice defects and deformations are produced in solids, and gaseous fission products migrate under the influence of the temperature gradient into hollow spaces formed in the central part of the solids. Volatile fission products may escape if there are leaks in the canning material. This makes continuous control of the activity in the coolant and purification of the latter by passage through ion exchangers necessary.

In order to compensate the excess reactivity, in water-cooled reactors, boric acid is added to the coolant in concentrations up to about 0.2%. The concentration is reduced with increasing burn-up. The pH is adjusted to ≈9 by the addition of 1−2 mg of 7 LiOH per liter of water to lower the solubility of the metal oxides and hydroxides, respectively, produced by corrosion on the walls of the cooling system.

## 12.3 Light Water Reactors

All light water reactors rely on the thermal fission of enriched uranium and on normal or "light" water for neutron moderation and heat transfer. These machines fall into two major categories of research reactors and power reactors. The research reactors are small, on the order of 1 MW of thermal power, and are optimized to provide intense neutron fluxes for the irradiation of samples. These reactors are fueled with a few kilograms of enriched uranium (20%−90% $^{235}U$) in fuel rods that are clad with a zirconium alloy or with aluminum. The entire core assembly of a 1 MW research reactor is on the order of 1 $m^3$ and can produce an internal neutron flux on the order of $10^{13}/cm^2 \cdot s$. Nuclear power reactors are generally much larger, on the order of 2 GW of thermal power, and are designed to produce electricity by the adiabatic expansion of steam in a turbine. There are two competing designs for nuclear power reactors that differ in the primary cooling loop. In one case the water is allowed to boil and in the other design super-heated water is held in the liquid phase under pressure.

Boiling water reactors are characterized by having only two coolant loops. The water in the primary coolant loop circulates through the reactor core and boils at approximately 1 atmosphere pressure and is heated to approximately 300 ℃. The steam is passed to a turbine system to generate electricity, is condensed, and is cycled back to the core. A second coolant loop is used to maintain a constant output temperature at the exit of the turbines; this loop removes the so-called waste heat at the end of the thermodynamic cycle. Such cooling loops are commonly included in machines that use adiabatic expansion to do work; for example, radiators are connected to gasoline engines in cars. The waste-heat loop in a nuclear power plant is usually an external, open loop. The waste heat is released in the atmosphere in large evaporative-cooling towers or released into rivers, lakes, or the ocean. The primary coolant is also the neutron moderator and is subject to intense irradiation in the core. It will contain radioactivities from impurities extracted from the walls, and so forth, and as a result the turbines will become contaminated. Thus, the important feature of the BWR design with the primary coolant circulating through the turbines necessitates placing them inside the containment shielding.

PWRs have two closed loops of water circulating in the plant plus a third, external loop to remove the waste heat. Water is pumped through the reactor core in the primary coolant loop to moderate the neutrons and to remove the heat from the core as in the BWR. However, the reactor vessel is pressurized so that the water does not boil. Steam is necessary to run the turbines, so the primary loop transfers the heat to a secondary loop. The water in the secondary loop is allowed to boil, producing steam that is isolated from both the core and the outside. The water in the primary loop usually contains boron (as boric acid $H_3BO_3$ about 0.025 M) to control the reactivity of the reactor. The steam in the secondary loop is allowed to expand and cool through a set of turbines as in the BWR; the cold steam condenses and is returned to the primary heat exchanger. A third loop of water is used to maintain the low-temperature end of the expansion near room temperature and remove the waste heat.

The PWR is more expensive to build because the reactor vessel must be stronger to withstand the higher water pressure, and there is a secondary coolant loop with pumps and so on. The BWR, while less expensive to build, is more complicated to service since the turbines are part of the primary coolant loop. The details of the core design are different as well. Approximately twice as many PWRs have been constructed as BWRs.

## 12.4 Fast Breeder Reactors

The principal design of a fast breeder reactor consists of a central core of plutonium in which fission occurs, surrounded by an outer blanket of $^{238}$U in which neutrons are captured to form new $^{239}$Pu. This blanket is surrounded by a reflector, usually of iron. The fission yield curve of $^{239}$Pu is similar to that of $^{235}$U but its heavy mass peak is shifted up by a few atomic mass units. Some neutron capture occurs in $^{239}$Pu yielding fertile $^{240}$Pu, which through another neutron capture produces fissile $^{241}$Pu; similarly, successively higher elements are formed in time. In a strong neutron flux their fate is destruction by fission. Thus, in the long run, by conversion of $^{238}$U into

$^{239}Pu$, etc., fission energy is always released and, in principle, all the $^{238}U$ can be used for producing fission energy. In practice a value of about 70% is considered more realistic. Still this would mean about 100 times more energy than is available from fission of $^{235}U$ only. Since the fast breeder reactor concept extends the fission energy resources by about a factor of 100, using a proved technology, it makes uranium the largest fossil energy resource presently available on earth. In addition fast reactors may also be used to fission long lived minor actinides such as americium and thereby not only increase the energy usage of the fuel but also shorten the longlevity of the final nuclear waste. Thus, if a separation process is designed to recover both uranium and plutonium together with the minor actinides a new fuel could be made directly using this composition. This is often called homogenous recycling.

The longest experience with fast research reactors has been provided by the BN-350 reactor in the former USSR, now Kazakhstan. It was operated at a power level up to 1 000 $MW_{th}$ from 1972 to 1994. Most of the heat was used for water desalination (120 000 $m^3$/d) and only a small amount for electricity production. Larger (100~1 200 $MW_e$) prototype fast breeder power reactors have been designed in the former USSR, France, the UK, and Japan. The largest prototype FBR built and operated so far is the Creys-Malville plant (located in France), which could produce 1 200 $MW_e$.

Figure 12.4 shows the Creys-Malville reactor in operation from 1986 to 1997. The thermal rating was 3 000 $MW_{th}$ (gross 1 240 $MW_e$); the neutron flux was very intense, $\leqslant 6 \times 10^{19}$ n $m^{-2}s^{-1}$, which puts a severe strain on construction materials. The core consisted of an inner and an outer fuel zone surrounded by a blanket, a steel reflector and a neutron shield, all immersed in liquid sodium (3 300 t) contained in a 25 cm thick austenitic steel vessel at atmospheric pressure. The reactor vessel was enclosed in a safety vessel and surrounded by a cooled concrete (biological) shield. The fuel was made of pins placed in hexagonal shrouds. The fuel zones contained 15 vol. % $PuO_2$ and 85 vol. % depleted $UO_2$, while the breeding blanket only contained depleted $UO_2$ (0.2% $^{235}U$). The fuel pins were clad in 0.5 mm stainless steel. The total plutonium amount was 4 800 kg. The neutron shield consisted of hollow steel tubes. Because the sodium becomes extremely radioactive due to $^{24}Na$ formation, the whole primary cooling system was contained in the reactor vessel. By means of a sodium/sodium intermediate heat exchanger within the reactor vessel (but outside the neutron flux) heat from the primary circuit (542 ℃) was transferred to a secondary sodium stream (525 ℃) and transported to a sodium/water steam generator which produced 487 ℃ steam at 17.7 MPa. Heat exchanger and piping were made of steel.

Creys-Malville (Superphenix). The vessel was filled with 3 300 t of liquid sodium.

The prompt neutron lifetime in a fast reactor is about 1 000 times smaller than in a thermal reactor. A reactor which can go critical on the prompt neutrons only would be exceedingly difficult to control. Therefore, fast reactors are designed to depend on the delayed neutrons (like thermal reactors). The time period ($t_{per}$) is large enough to allow reactor control through the use of neutron absorbing rods. Since the neutron spectrum is such that several percent of the flux is in the resonance region, control rods with boron can be used; in practice boron carbide (the carbon

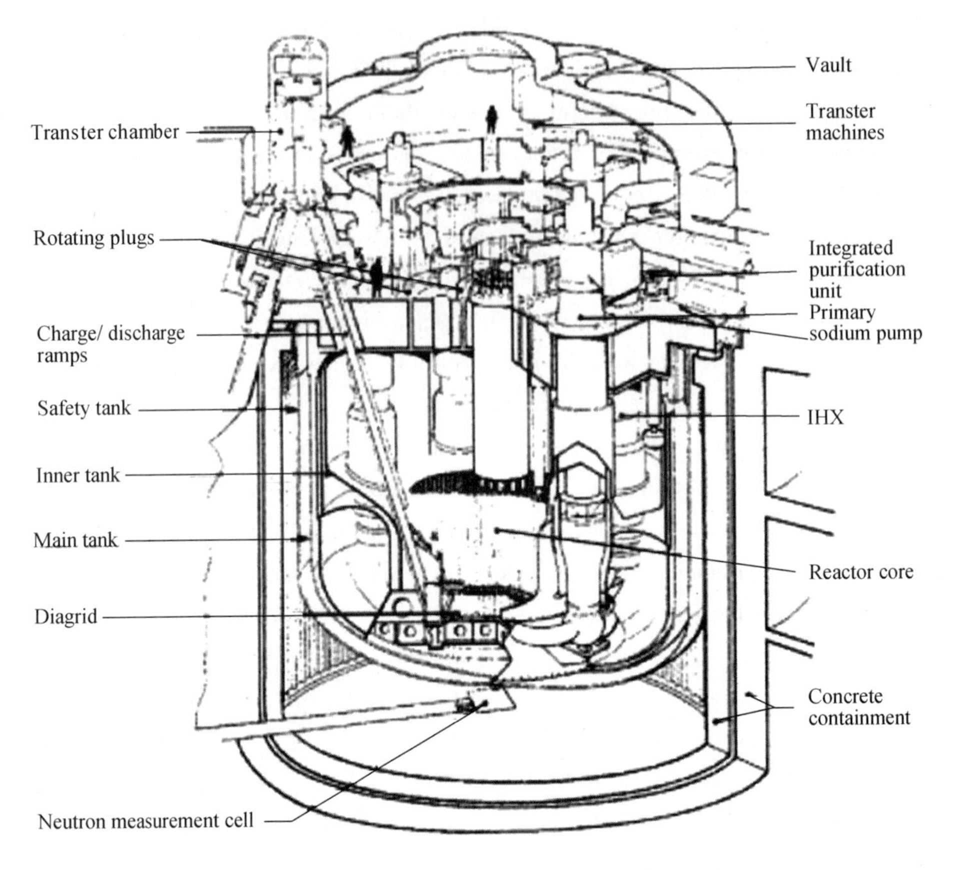


**Figure 12.4 Vertical section of European prototype fast power reactor**

atom reduces the neutron energy further) and/or tantalum (which has large absorption peaks ≥ $10^4$ b at 3.00 eV) are used.

Although the temperature coefficient of the Creys-Malville reactor was negative, this is not so for fast Pu-fuelled reactors with harder neutron spectra, e. g. the Russian BN-350. If the fast reactor becomes overheated, the core could be deformed, making it prompt critical. The power would then increase rapidly with a doubling time in the microsecond range, and a severe accident would be unavoidable. To prevent this the core is designed to achieve negative reactivity upon sudden power transients by using $^{238}UO_2$ in the core. When the temperature rises, doppler broadening occurs in the $^{238}U$ resonance capture region, and consequently more neutrons are consumed by $^{238}U$, which limits the power increase. The shorter the neutron lifetime, the less would be the power excursion.

Considerable radioactivity is induced in the sodium in the primary cooling circuit of a liquid metal cooled fast breeder reactor (LMFBR): $^{23}Na(n, \gamma)$ $^{24}Na$, $^{23}Na(n, p)$ $^{23}Ne$, etc. $^{24}Na$ has a 15 h $t_{1/2}$ and emits energetic γ's. The primary cooling loop must therefore be well shielded. Its activity is a nuisance only in case of repair work in the primary system, requiring considerable waiting time before the loop can be approached. The sodium dissolves many of the corrosion and fission products eventually released. To remove these, the primary loop is provided with cold trap purification systems.

### 词汇

| | |
|---|---|
| enriched U | 浓缩铀 |
| moderator | 慢化剂 |
| homogeneous | 均相的 |
| heterogeneous | 非均相的 |
| molten salt | 熔盐 |
| converter | 转换器 |
| ship propulsion | 船舶推进 |
| concrete | 混凝土 |
| isotope separation | 同位素分离 |
| collision | 碰撞 |
| thermal conductivity | 热导率 |
| pumping power | 泵功率 |
| precaution | 预防 |
| specific ionization | 比电离 |
| lattice defect | 晶格缺陷 |
| migrate | 迁移 |
| volatile | 挥发性的 |
| ion exchanger | 离子交换剂 |
| neutron flux | 中子通量 |
| adiabatic expansion | 绝热膨胀 |
| primary cooling loop | 主冷却回路 |
| waste-heat loop | 废热回路 |
| evaporative-cooling tower | 蒸发冷却塔 |
| immerse | 浸没 |
| delayed neutron | 缓发中子 |
| resonance region | 共振区域 |
| power excursion | 功率激增 |
| prompt neutron | 瞬发中子 |

### 注释

1. The next nuclear reactor began operation in 1943 at Oak Ridge (USA) by using 54 t of uranium metal in the form of fuel rods inserted into a block of graphite 5.6m long, shielded by concrete.

下一个核反应堆于 1943 年在美国橡树岭开始运行,它将 54 吨铀金属以燃料棒的形式插入 5.6 米长的石墨块中,并用混凝土屏蔽。

2. They exhibit several advantages: the high burn-up is due to the fact that large amounts of $^{238}U$ are transformed into the easily fissile $^{239}Pu$. In this way, the energy production from U is

increased by a factor of about 100, and enrichment of $^{235}U$ by isotope separation is not needed.

它们有几个优点:由于大量的$^{238}U$转化为易裂变的$^{239}Pu$,所以燃耗高。这样,铀产生的能量增加了约 100 倍,并且不需要通过同位素分离来浓缩集$^{235}U$。

3. The purpose of the moderator is to take away the energy of the fission neutrons by collisions, without absorbing appreciable amounts of the neutrons.

使用慢化剂的目的是通过碰撞带走裂变中子的能量,而不吸收相当数量的中子。

4. All other materials used in nuclear reactors for construction or as tubes should exhibit low neutron absorption, low activation, no change in properties under the influence of the high neutron and γ-ray fluxes, and high corrosion resistance.

在核反应堆中用于建造或作为管的所有其他材料应具有低中子吸收、低活性,在高中子和 γ 射线通量的影响下性能不发生变化,且具有高耐腐蚀性能。

5. Furthermore, lattice defects and deformations are produced in solids, and gaseous fission products migrate under the influence of the temperature gradient into hollow spaces formed in the central part of the solids.

此外,在固相中会产生晶格缺陷和变形,气体裂变产物在温度梯度的影响下会迁移到固相中心部分形成的空心中。

6. There are two competing designs for nuclear power reactors that differ in the primary cooling loop. In one case the water is allowed to boil and in the other design super-heated water is held in the liquid phase under pressure.

有两种相互竞争的核反应堆设计,它们在主冷却回路上有所不同。在一种情况下,水允许沸腾,而在另一种设计中,过热的水在压力下保持在液相中。

7. Since the fast breeder reactor concept extends the fission energy resources by about a factor of 100, using a proved technology, it makes uranium the largest fossil energy resource presently available on earth.

由于快中子增殖反应堆的概念使用一种经过验证的技术,将裂变能源资源扩大了大约 100 倍,它使铀成为地球上目前可用的最大的化石能源资源。

8. The prompt neutron lifetime in a fast reactor is about 1 000 times smaller than in a thermal reactor. A reactor which can go critical on the prompt neutrons only would be exceedingly difficult to control. Therefore, fast reactors are designed to depend on the delayed neutrons (like thermal reactors).

快堆中的瞬发中子的寿命大约比热堆小 1 000 倍。一个仅能在瞬发中子上达到临界的反应堆将是极其难以控制的。因此,快堆的设计依赖于延迟中子(如热堆)。

# Chapter 13 Nuclear Materials

## 13.1 Introduction

The nuclear industry requires a selection of materials having outstanding properties in various and sometimes very specific environments. In addition, irradiation leads to a selection of alloys with specific properties with respect to radiation level which induces alloy transformation (evolution of microstructures, of mechanical characteristics, etc.) and compatibility with neutron flux. Materials used in the nuclear industry may be divided into three main categories.

The first category includes materials that have been specifically developed for nuclear applications. They include fuel materials and alloys used for their specific nuclear neutron absorbing properties. Most of them are used in the reactor cores. They include the fissile materials (uranium, plutonium, oxides, carbides, etc.), where chain reactions are produced, and the neutron absorbers (silver, indium and cadmium-based alloys, boron, etc.) which are used to control the chain reaction. The behavior of these materials is dependent mainly on the irradiation process and on their evolution under irradiation. These materials will not be detailed here.

The second category comprises the few specific alloys that have been or are being developed for nuclear applications, but are not used only in the nuclear industry. It includes graphite (used as a moderator in gas-cooled reactors), zirconium alloys (even if zirconium alloys without hafnium are only used as clad materials in nuclear power plants due to their low capture neutron cross sections), oxide dispersion strengthened ferritic-martensitic steel which are developed for cladding of Generation Ⅳ reactors, or vanadium alloys for fusion systems.

The third category includes standard alloys (stainless steel, aluminum alloys, nickel-based alloys, etc.) which are used not only in nuclear environments but also in numerous other industrial environments. These alloys are used in systems which are not specific to nuclear facilities (e.g., alloys used in cooling circuits). More often, these alloys have not been specifically developed for the nuclear industry even if sometimes those alloys have been optimized for nuclear use. They are used widely (e.g., nickel base alloys in steam generators) as structural or reactor core materials even if they have to face irradiation damage.

In all nuclear systems, component materials are subject to various degradation processes, driven by applied loadings (radiation, thermal, mechanical, chemical). The irradiation itself may induce major damage which leads to major evolution in the microstructure, properties and behavior of irradiated materials. Irradiation damage includes hardening (reduction of ductility), chemical composition changes (e.g., helium formation inside alloys, grain boundary segregation/enrichment in some alloyed elements), swelling and deformation. These phenomena are very important and often determine the use of an alloy under irradiation conditions.

## 13.2 Zirconium Alloys

Zirconium alloys have been chosen for various nuclear applications due to several properties: low thermal neutron capture cross section together with good behavior in high temperature water were probably the main reasons for their use as clad materials in light water reactors, while the very high resistance to corrosion in specific nitric acid conditions together with good mechanical properties were needed for some apparatus (such as fuel dissolvers and nitric acid recovery concentrators) of reprocessing plants with the PUREX process.

Pure zirconium crystalizes in the alpha phase at ambient temperature (hexagonal close packed system) at room temperature and in the beta phase (body centreed cubic system) at high temperatures, transition phases occur at 864 ℃ for pure zirconium.

The zirconium alloys for nuclear applications are quite limited today: they are based either on a combination of tin (1%-2%) and iron additions or on niobium addition (1%-3%) for core use while unalloyed zirconium is used in reprocessing plants. The oxygen increases the yield strength. It is introduced intentionally as oxide powder up to a concentration of 0.5%-0.7 at.%. Tin was originally added at a concentration of 1.2%-1.7% to increase corrosion resistance. Zircaloys are Zr-Sn alloys. It contributes also to increase the mechanical properties, mainly the creep resistance of the alloys. Iron and chromium have been added (a few tenths of a percent) to improve mechanical properties and corrosion resistance due to a precipitation of Laves phases.

Because of their importance for neutronic properties of these alloys, it is a mayor issue to have a strict control of Co, Hf and U concentrations. Niobium is the main element of the second series of industrial zirconium alloys. Zr-Nb alloys were originally developed for Canadian CANDU reactors and for the cladding of Russian VVER reactors. These alloys are now developed for PWR and BWR applications. Niobium is soluble in the beta phase and allows a better control of the alpha-beta transformation during heat treatments.

Zirconium has a very high affinity for oxygen which creates a difficulty for industrial processing of Zr and Zr alloys. So, in contact with water or air, it is immediately covered with a protective oxide layer (passive film) which gives an excellent resistance to corrosion in many media considered as aggressive, notably in high acidic or basic ones. Its resistance to corrosion makes it a choice material for improving performance or saving on maintenance operations. In that context, the case of nitric acid, and especially nitric media encountered in reprocessing plants, may serve a model: zirconium is used in numerous situations encountered when stainless steels are not resistant enough to corrosion.

## 13.3 Stainless Steels

Type 304 and 316 austenitic stainless steels are the main stainless steels used in nuclear facilities including reprocessing plants and nuclear power reactors: they are the alloys of pressure boundary piping of BWRs and of the primary circuits of PWRs. The internal surfaces of pressure vessels are also cladded with 308/309 stainless steel weld overlays. They are also extensively used in sodium fast-cooled reactors. In VVER, austenitic steels are also widely used, including the tubes of steam generators which are made from an austenitic stainless steel (18% Cr, 10% Ni stabilized with titanium). The VVER pressure vessels are also cladded internally with two stainless steel weld layers. The inner layer is non-stabilized stainless steel and that in contact with the coolant is a niobium-stablilized stainless steel. The use of these stainless steels in nuclear facilities is due to their excellent resistance to uniform corrosion and to their good mechanical properties at operating temperatures. Low carbon grades of austenitic stainless steels are now used to minimize the risk of sensitization by grain boundary chromium depletion due to chromium carbide precipitation in weld heat affected zones. Nevertheless, such sensitized materials are present in old reactors. Practical experience shows that in reducing environments like in hydrogenated PWR primary water, intergranular stress corrosion cracking does not occur in such sensitized materials, in contrast to BWRs with oxygenated normal water chemistry.

However, in nuclear power plants, alloys close to core are subject to high-energy neutron fluxes. Neutron radiation causes atom placements creating vacancies and interstitials (defects) which either recombine or diffuse to traps such as dislocations, grain boundaries and precipitate interfaces. After neutron irradiation, the initial dislocation network present in cold worked austenitic stainless steels is removed and a new dislocation microstructure is observed leading to a significant hardening of materials with yield stresses typically in the range of 800 – 1 100 MPa. Defect trapping at grain boundaries leads to changes in elemental composition of these grain boundaries, phenomena known as radiation-induced segregation: typically enrichment has been observed for nickel and by factors up to 3 to 10 for silicon and phosphorus, while depletion is observed for chromium, iron and molybdenum. These modifications of grain boundary chemical composition, particularly the chromium depletion, have been shown to be important for the initiation and the propagation of irradiation-assisted stress corrosion cracking (IASCC) in oxidizing environments (BWRs), while it is mainly the hardening due to neutron irradiation that is the mayor parameter leading to IASCC in reducing environments (PWRs). In addition to hardening and grain boundary segregation, other irradiation damage occurs on austenitic stainless steel: changes in elemental composition may also be caused by nuclear transmutation reactions. This is the case for helium [nickel allows in-situ formation of helium by (n, $\alpha$) reactions] and hydrogen by (n, p) reactions, which may form bubbles and may have an adverse effect on IASCC resistance.

In the sodium-cooled fast reactors, austenitic stainless steels (316 type) are used as material for fuel assemblies including the fuel cladding tubes (zirconium cannot be used as in light water reactors), thanks to a low capture cross section of these alloys for fast neutron spectrum. But microstructural evolution and significant swelling occur above 30 dpa. This swelling phenomenon is a critical issue, since it affects directly the geometry and performance of the core of sodium-cooled reactors. Major developments have been performed to minimize the swelling of stainless steels with new materials like 15/15 Ti stainless steels (15% Cr and 15% Ni), with slight modifications of minor elements such as Si and P, and also with martensitic 90Cr steel (wrapper applications) and oxide dispersion strengthened (ODS) steels with 9%-17% chromium (ferritic-martensitic steels) for fuel pin cladding applications. Due to activation problems, the Co content us limited for those steels used for SFR core materials.

High strength stainless steels are also widely used in water reactors for components such as bolts, springs or valve systems. The main ones are martensitic stainless steels such as A410 and 17-4PH, and A286 precipitation hardened austenitic steels. A small number of them have cracked due to stress corrosion cracking or hydrogen embrittlement. In most cases, failures have been associated with hardness values too high which have been aggravated by impurities. High strength stainless steel used in primary water components like valves trim, for instance, could be susceptible to SCC in primary water or in concentrated boric acid solutions. Susceptibility of these steels grades are highly influenced by the in-service aging of the martensitic phase which are intended to occur when low initial aging temperatures are selected. Adequate thermal treatment conditions were derived from field experience in PWRs.

Cold-worked austenitic stainless steels can be susceptible to SCC in BWR and PWR conditions. SCC of cold-worked austenitic stainless steel in BWR is a well-known phenomenon, but SCC of cold-worked austenitic stainless steels in the reducing environment of PWRs as only recently observed, due to large initiation times and high values of cold work required for the cracking to occur.

## 13.4 Nickel Alloys

Nickel alloys have been selected for steam generator (SG) tubes mainly due to their good resistance to stress corrosion cracking (compared to stainless steels), along with high mechanical properties. Ni alloys, also known as Inconel, which contain 15% Cr (Alloy 600), or 30% Cr (Alloys 690) are mainly used for SG tubes in PWRs, some reactor pressure vessel top and bottom heads and cooling system welds (Alloys 182, 82, 52 and 152). Alloy 800 (Incoloy 800), which is not a Ni-based alloy but contains 33% Ni and 22% Cr with 40%-45% Fe, is also used mainly for SG tubes. In fact, Alloys 600 and 182, first selected in the US for SG tubes during 1960s, have proved to be very susceptible to SG cracking in primary and secondary waters of pressurized reactors. Quite rapid cracking of elements such as SG tubes, pressurized vessel head penetrations and pressurizer nozzles have been observed. These cracks and leaks lead to repairs (pressurizer nozzles) or replacements (SGs, RPV upper head). The replacement alloys (690, 52, 152) are much more resistant to SCC, probably because of their lower Ni content, Alloy 800 is immune to SCC in primary water conditions and highly resistant to SCC in common secondary conditions as it is clearly reflected by steam generator field experience gained in Germany (PWRs) and Canada (CANDU).

High Strength Inconels X750 and X718 are used for internal components (e. g., for guide tube support pins and clad spacer grids, respectively). Susceptibility of X750 to SCC is similar to that of Alloy 600. Remedies are essentially based on component design (design stresses) and improvement of fabrication steps (thermal treatment): Inconel 718 is sparsely susceptible to SCC and a strong heat to heat dependency is observed.

Ni-based alloys have traditionally also been used for high temperature applications in many industries. So, in early high temperature reactors cooled with helium (HTR), mainly Alloy 800 and chromium-rich nickel alloys reinforced with Co and/or Mo (e. g., Alloy 617) have been used. For Gen-Ⅳ reactors such as gas-cooled reactors (very high temperature reactors, VHTR), structures will be exposed to temperatures considerably higher than previously for HTR. Other grades are assumed a priori to have better performance at high temperature, particularly owing to their optimized creep resistance [e. g. Cr-rich nickel alloys reinforced with tungsten, typically, Alloys 230 and 740, oxide dispersion strengthen (ODS) nickel alloys]. These new Ni-base superalloys have good creep rupture properties and high temperature strength. The main problem with Ni-base alloys are the radiation embrittlement, swelling and phase instability under the neutron radiation environment. Their applicability is more probable in components (e. g., turbines, steam generators) where radiation effects are minimal.

## 13.5 Other Alloys

### Cobalt alloys

Cobalt base alloys, often called Stellite, are very wear resistant and are the materials of choice for some specific parts where have good resistance to various forms of wear and high strength over a wide range of temperatures are needed, like valves (valve seat hard-facing), for instance. Stellite 6, the nominal chemical composition of which is 27%-32% Cr, 4%-6% W, 0.9%-1.4% C, with additions of Ni, Fe, Si, Mn and Mo and Co as balance, is a two-phase alloy, with a chromium carbide minor phase, which has been shown to have outstanding resistance to both erosion and corrosion. These unique properties have led to its use in light water nuclear reactor components subject to unusually severe wear/corrosion conditions. The good corrosion behavior in high light water reactor environments is mainly due to the high chromium content of the alloy and linked to a strongly enhanced chromium concentration within the oxide film. The Stellite 6 generally selected in PWR/BWR is highly resistant to corrosion in standard conditions and most of the degradation was connected to excessive applied mechanical stresses or fabrication defects. Even if the uniform corrosion of these alloys in aqueous environment is low, nevertheless some Co is released into the circulating water and the $^{59}$Co (the only natural isotope of cobalt) is activated by neutron flux in the core of the reactor to become $^{60}$Co, a radioisotope with a half-life of five years and which releases energetic gamma rays. So the replacement of cobalt alloys in the core circuit of nuclear power plant is priority to minimize the contamination of the core circuit and so the exposure of nuclear worker. Replacement of stellite by iron base alloys have already been performed in some nuclear power plants and is planned in Generation Ⅲ and Ⅳ power plants.

**Copper and copper alloys**

These are widely used in condensers for tubes (admiralty brass, 90%-10% and 70%-30% copper nickel alloys) or for plates (aluminum brass). Corrosion of copper alloy tubes has been widely reported and today stainless steels and titanium are chosen respectively for river and sea water exposure. Copper release rate in secondary water is highly susceptible to transient condition and transport of copper ions in SG has strong detrimental consequences on the resistance of SG tubes to localized corrosion and SCC in secondary environments. Erosion corrosion of copper alloys was also observed on the secondary side of condenser tubes in relation with flow regimes encountered during vapor condensation.

Titanium (unalloyed Ti is selected for condensers) was shown to be susceptible to erosion corrosion by droplets during vapor condensation and remedies are strictly based on improvement of condenser design.

Copper is planned also for the overpacks of highly radioactive nuclear waste in Swedish and Finnish concepts of geological waste disposal ingranitic rocks: in geological granitic formations, water is not aerated under these non-oxidizing conditions, copper is thermodynamically stable and so may survive for millennia without corrosion.

**Aluminum alloys**

These are mainly used in experimental nuclear reactors, as cladding material and also as structural material. Test reactors are made to produce neutrons rather power; they require fuels that can operate at high-power density (therefore, high cladding surface heat flux) with low parasitic neutron absorption. Consequently, several materials have been used as cladding materials, including commercially pure aluminum 2S (currently called Alloy 1100), AG3NE alloy cladding (3% Mg, 97% Al), Alloy Al 6061 also called AlFeNi alloy and mainly composed of aluminum (about 96 wt%), iron (1%), nickel (1%) and magnesium (1%). After being irradiated for more than 50 a in the BRI reactor at low temperature, the natural-uranium-aluminum (pure aluminum 2S) clad fuel rods are found to be in very good condition. The AlFeNi alloy was developed during the 1950s for nuclear fuel clad application. The main goal was to increase the corrosion resistance of the material at high temperature (between 250 and 300 ℃) for high flux reactors without decreasing the neutron adsorption performance of aluminum. It was developed on the basis of its good corrosion resistance in water at high temperature. It is already used as fuel cladding in two other experimental reactors and has been chosen for the new Jules Horowitz material testing reactor, for the cladding of the nuclear fuel plates.

In addition to specific irradiation conditions, aluminum alloys used for piping and structural components can be susceptible to galvanic or crevice corrosion. Dedicated design and strict control of the water purity are prerequisites.

**词汇**

transformation　　变形

| | |
|---|---|
| mechanical characteristics | 机械特性 |
| compatibility | 兼容性 |
| neutron absorber | 中子吸收剂 |
| clad material | 包壳材料 |
| dispersion | 扩散 |
| ferritic-martensitic steel | 铁素体-马氏体不锈钢 |
| steam generator | 蒸汽发生器 |
| hardening | 硬化 |
| ductility | 延展性 |
| grain boundary segregation | 晶界分离 |
| swelling | 肿胀 |
| deformation | 变形 |
| hexagonal close packed system | 六边形紧密堆积体系 |
| body centreed cubic system | 体心立方体系 |
| yield strength | 屈服强度 |
| creep resistance | 抗蠕变性 |
| passive film | 钝化膜 |
| pressure boundary | 压力边界 |
| primary circuit | 主回路 |
| pressure vessel | 压力容器 |
| sensitization | 敏化 |
| chromium depletion | 贫铬 |
| intergranular stress corrosion | 晶间应力腐蚀 |
| hydrogen embrittlement | 氢脆 |
| head penetration | 封头贯穿件 |
| pressurizer nozzle | 稳压器管口 |
| guide tube support pin | 导管支撑销 |
| clad spacer grid | 包壳间格架 |
| creep rupture | 蠕变断裂 |
| erosion | 侵蚀 |
| fabrication defect | 制造缺陷 |
| condenser | 冷凝器 |
| admiralty brass | 海军黄铜 |
| transient condition | 瞬时工况 |
| granitic rock | 花岗岩石 |
| galvanic corrosion | 电偶腐蚀 |
| crevice corrosion | 缝隙腐蚀 |

## 注释

1. They include the fissile materials (uranium, plutonium, oxides, carbides, etc.), where

chain reactions are produced, and the neutron absorbers (silver, indium and cadmium-based alloys, boron, etc.) which are used to control the chain reaction. The behavior of these materials is dependent mainly on the irradiation process and on their evolution under irradiation.

它们包括链式反应产生的易裂变材料(铀、钚、氧化物、碳化物等),以及用来控制链式反应的中子吸收剂(银、铟及铬基体合金、硼等)。这些材料的行为主要取决于辐照过程及其在辐照下的演变。

2. It includes graphite (used as a moderator in gas-cooled reactors), zirconium alloys (even if zirconium alloys without hafnium are only used as clad materials in nuclear power plants due to their low capture neutron cross sections), oxide dispersion strengthened ferritic-martensitic steel which are developed for cladding of Generation Ⅳ reactors, or vanadium alloys for fusion systems.

它包括石墨(用作高温气冷堆中的慢化剂)、锆合金(即使不含铪的锆合金,由于中子俘获截面低,只用作核电站中的包壳材料),用于第四代反应堆包壳的氧化物弥散强化铁素体-马素体不锈钢,或用于聚变系统的钒合金。

3. In all nuclear systems, component materials are subject to various degradation processes, driven by applied loadings (radiation, thermal, mechanical, chemical). The irradiation itself may induce major damage which leads to major evolution in the microstructure, properties and behavior of irradiated materials. Irradiation damage includes hardening (reduction of ductility), chemical composition changes (e.g., helium formation inside alloys, grain boundary segregation/enrichment in some alloyed elements), swelling and deformation. These phenomena are very important and often determine the use of an alloy under irradiation conditions.

在所有核系统中,组件材料都受到应用负载(辐照、热力、机械、化学)驱动的降解过程。辐照本身可能诱发重大损伤,导致受照材料的微观结构、性质和行为发生明显变化。辐照损伤包括硬化(延展性降低)、化学成分变化(例如,合金内部生成氦,一些合金元素晶界分离/浓缩集)、肿胀及变形。这些现象非常重要,常常决定了合金在辐照条件下的使用。

4. Zirconium has a very high affinity for oxygen which creates a difficulty for industrial processing of Zr and Zr alloys. So, in contact with water or air, it is immediately covered with a protective oxide layer (passive film) which gives an excellent resistance to corrosion in many media considered as aggressive, notably in high acidic or basic ones. Its resistance to corrosion makes it a choice material for improving performance or saving on maintenance operations.

锆对氧有很强的亲和性,这给锆及锆合金的工业加工带来了困难。因此,在与水或空气接触时,它很快会被一层保护性氧化层(钝化膜)所覆盖,钝化膜在许多腐蚀性介质中具有优异的耐腐蚀性能,特别是在高酸或碱性介质中。它的耐腐蚀性使其成为提高性能或节省维修操作的选择材料。

5. Low carbon grades of austenitic stainless steels are now used to minimize the risk of sensitization by grain boundary chromium depletion due to chromium carbide precipitation in weld heat affected zones. Nevertheless, such sensitized materials are present in old reactors. Practical experience shows that in reducing environments like in hydrogenated PWR primary water, intergranular stress corrosion cracking does not occur in such sensitized materials, in contrast to BWRs with oxygenated normal water chemistry.

低碳等级的奥氏体不锈钢现在被用于减小因焊缝热影响区析出碳化铬而导致晶界铬耗损的敏化风险。然而,这种敏化材料存在于旧反应堆中。实践经验表明,在像氢化压水堆原水这样的还原性环境中,与含氧正常水化学反应的沸水堆相比,这种敏化材料不会发生晶间应力腐蚀开裂。

# Chapter 14 Radiation Chemistry

## 14.1 Energy Transfer

It was learned that radiation could cause changes in matter. Glass ampoules containing milligrams of radium darkened within a few months and become severely cracked, allowing the leakage of radon gas. Among the radiation effects observed were the fuorescence induced in different salts and the changes in their crystallographic form resulting in color changes from delocalised charges in the crystals. Metals were found to lose their elasticity and became brittle due to delocalised atoms breaking the inherent structure. Radiation was also found to have a profound effect on the chemical composition of solutions and gases. Water, ammonia and simple organic substances decomposed into more elementary constituents and also combined into more complex polymeric products. Radiation decomposition (radiolysis) of water caused evolution of hydrogen and oxygen gas and formation of hydrogen peroxide.

The chemical effects of radiation depend on the composition of matter and the amount of energy deposited by the radiation. High energy radiation can be divided into (1) charged particles and (2) uncharged particles (n) and electromagnetic radiation. The latter produce recoil atomic ions, products of nuclear reactions and electrons as charged secondary ionizing particles. The terms direct and indirect ionizing radiation are often used for (1) and (2) respectively.

The energy of charged particles is absorbed mainly through ionization and atomic excitation. For positrons the annihilation process must be considered. For electrons of high kinetic energy bremsstrahlung must be taken into account. The average energy, $w$, for the formation of an ion pair in gaseous material by charged particles is between 25 eV and 40 eV. For the same absorbing material, it is fairly independent of the type of radiation and the energy. While it is easy to measure $w$ in a gas, it is more difficult to obtain reliable values for liquids and solids. They also differ more widely; e.g., $w$ is 1 300 eV per ion pair in hexane (for high energy electrons) while it is about 5 eV per ion pair in inorganic solids.

Another important concept is the linear energy transfer (abbreviated as LET) of charged particles. It is defined as the energy absorbed in matter per unit path length travelled by a charged particle. For the same energy and the same absorbing material, the LET values increase in the order: high energy electrons (also approximatively $\gamma$-rays) < $\beta$-particles (also approximately soft X-rays) < protons < deuterons < heavy ions (ions of N, O, etc.) < fission fragments.

When neutrons or photons having the incident particle energy $E_{in}$ are absorbed, a certain fraction of energy $E_{tr}$ is transferred into kinetic energy of charged particles when traversing the distance dx. We can define an energy transfer coefficient as

$$\mu_{tr} = E_{in}^{-1} dE_{tr}/dx (m^{-1}) \tag{14.1}$$

In radiation chemistry, the $G$ value is used as a measure of the chemical yield. It is defined by the number of molecules formed or decomposed per 100 eV of energy absorbed in the system. For example, $G(H_2) = 3$ means that 3 molecules of $H_2$ are formed per 100 eV absorbed and $G(-H_2O) = 11$ means that 11 molecules $H_2O$ are decomposed per 100 eV absorbed. Some examples of radiation-induced reactions are given in Table 14. 1. Chain reactions exhibiting $G$ values $>10^3$ are of greatest interest for technical applications.

**Table 14. 1 Examples of radiation-induced reactions**

| Reactions | $G$ values |
|---|---|
| Production of $O_3$ by irradiation of $O_2$ | 6–10 |
| Production of $NO_2$ by irradiation of $N_2/O_2$ mixtures (by-products NO and $N_2O$) | 1–7 |
| Production of $C_2H_5Br$ by irradiation of a mixture of $C_2H_4$ and HBr | $>10^5$ |
| Production of $C_2H_5Cl$ by irradiation of a mixture of $C_2H_4$ and HCl | $\approx 10^4$ |
| Chlorination (similar to the photochemical reaction by UV), for example, by irradiation of a mixture of $C_6H_6$ and $Cl_2$ | $10^4$–$10^5$ |
| Oxidation of carbohydrates; production of phenol by irradiation of a mixture of $C_6H_6$ and $O_2$ | $10^4$–$10^5$ |
| Sulfochlorination (similar to the photochemical reaction by UV), for example, production of sulfonic acid chlorides by irradiation of mixtures of carbohydrates, $SO_2$, and $Cl_2$ | $\approx 10^7$ |
| Production of alkylsulfonic acids by irradiation of mixtures of carbohydrates, $SO_2$, and $O_2$ | $10^3$–$10^4$ |

## 14. 2 Radiation of Gases

Radiochemists are mainly concerned with the effect on air during the storage of radioelements. Because of their low density, the primary species are not confined to the particle track, but diffuse rapidly throughout the gas. Accordingly, LET effects are reduced.

The main chemical compounds that appear in irradiated air are ozone, easily detected by its odor even at very low concentrations, and nitrous oxides. The radiation stability of many compounds in aqueous solutions depends to a large extent on the presence of dissolved oxygen, owing to the formation of peroxides. In such cases, solutions of radioactive matter must be stored under air-free conditions. In any event, the free space above these solutions is rapidly occupied by $H_2$, $O_2$, $CO_2$, and other gaseous radiolytic degradation products.

Radioactive matter containing the radionuclides $^{226}Ra$, $^{224}Ra$, and $^{223}Ra$ continuously generate short-lived isotopes of the gaseous element radon. The emanation power of the medium is the fraction of the formed radon atoms that are released into the surroundings. It may reach 100% for solid organic compounds or in heated solutions of the precursors.

The longest-lived radon isotope, $^{222}Rn$, has been extensively used for investigation of the chemical effects of α particles in gaseous systems. It decays successively to ephemeral isotopes of the solid elements Po, Bi, and Pb, which are adsorbed on aerosols and dust particles or deposited on the walls of the containers. When these species are released into the air, they constitute a radiological hazard because they can be inhaled and retained by the broncho-respiratory tract. Eventually, the long-lived daughters $^{210}Pb$ and $^{210}Po$ grow in and produce a durable radioactive contamination of the air in addition to the effects of radiolysis.

## 14.3 Radiation of Aqueous Solutions

In radiochemistry, the most common forms of radioactive matter are probably those in aqueous solutions. In dilute solutions below $10^{-3}$ M, the effect of radiation is essentially indirect. The energy of the radiation is absorbed in the solvent. Along the path of the ionizing particles, water molecules are either ionized,

$$H_2O \longrightarrow H_2O^+ + e^- \quad (14.2)$$

or excited,

$$H_2O \longrightarrow H_2O^* \quad (14.3)$$

The ionized water molecule rapidly undergoes the ion-molecule reaction

$$H_2O^+ + H_2O \longrightarrow H_3O^+ + {}^\circ OH \quad (14.4)$$

with simultaneous formation of the free radical $^\circ OH$. Here $H_3O^+$ stands for the solvated proton and the dot on $^\circ OH$ and on other symbols represents the single electron of a free radical. The electron released in the ionization of a water molecule generally has enough kinetic energy to escape from the attraction sphere of the positive molecular ion and diffuses in the medium. It becomes thermalized in about 1 ps and eventually forms an aquated species, the hydrated electron:

$$e^-\ (\text{fast}) \longrightarrow e^-\ (\text{thermal}) \longrightarrow e^-_{aq} \quad (14.5)$$

The excited water molecule decomposes into two free radicals:

$$H_2O^* \longrightarrow {}^\circ H + {}^\circ OH \quad (14.6)$$

The three species $^\circ OH$, $H^\circ$ and $e^-_{aq}$ are initially localized in the tracks. Their further fate is governed by competition between diffusion out of the tracks and very fast mutual reactions which lead to the formation of molecular products:

$$^\circ H + {}^\circ H \longrightarrow H_2 \quad (14.7)$$

$$^\circ OH + {}^\circ OH \longrightarrow H_2O_2 \quad (14.8)$$

$$^\circ H + e^-_{aq} \longrightarrow H_2 + OH^- \quad (14.9)$$

$$e^-_{aq} + e^-_{aq} \longrightarrow H_2 + 2OH^- \quad (14.10)$$

Including the back reaction

$$^\circ OH + {}^\circ OH \longrightarrow H_2O \quad (14.11)$$

For particles with high LET, such as ex-particles and fission fragments, the density of primary species is very high and mutual reactions are favored. For fast electrons on the other hand, more free radicals diffuse into the bulk of the solutions.

The dependence of radiochemical yields of free radicals and molecular products in neutral water as a function of the LET is shown in Figure 14.1. For low LET radiation, yields amount to 2.7 for the aqueous electron, 0.6 for $^\circ H$ radicals, and 2.7 for $^\circ OH$ radicals; the radiochemical yields are 0.45 for $H_2$ and 0.75 for $H_2O_2$. The relative yields of the reducing species $^{\circ}H$ and hydrated electron are pH-dependent because of the reaction

$$e^-_{aq} + H_3O^+ \longrightarrow {}^{\circ}H + H_2O \tag{14.12}$$

In the presence of air, $^{\circ}H$ is converted to the hydroperoxide radical $HO_2^{\circ}$ and $e^-_{aq}$ to $O_2^-$; the overall yield of reducing agents decreases.

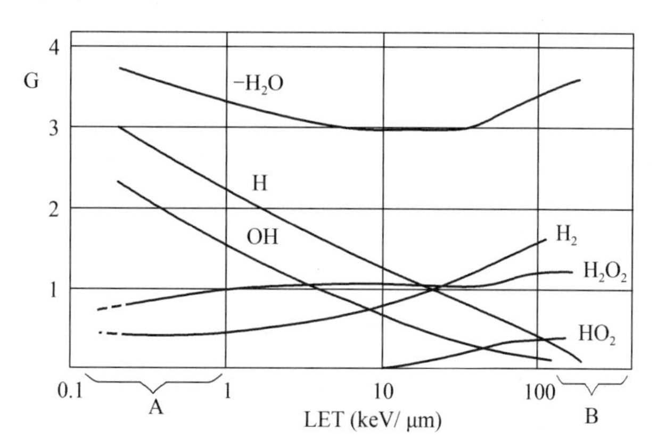


**Figure 14.1 Radiochemical yields of the radicalar and molecular products formed in the radiolysis of aerated neutral water, as a function of the LET of the incoming radiation. (A) Electrons up to 3 MeV and 1 MeV gamma rays; (B) 5-MeV α particles.**

## 14.4 Radiation of Solid Compounds

Radiation effects of solid compounds were first observed in the coloration of glassware in contact with radioelements, and in pleochroic halos in mica irradiated by the inclusion of radioactive substances. Another consequence of radiation damage in radioactive minerals is metamictization, i.e., the conversion of a crystalline state into an amorphous one. A concentration of U or Th as low as 1% is sufficient to cause complete metamictization of some minerals in the course of geological time.

Self-radiation effects in solids are mainly due to the displacement of atoms by α particles, recoil nuclei, and neutrons. These effects are very important when one is dealing with macroscopic amounts of short-lived radionuclides. In a $^{210}Po$ compound, each atom is displaced several times per day under the combined effects of particles and recoiling $^{206}Pb$ atoms. The chemical effects of self-irradiation depend on the type of compounds. $^{239}PuF_6$ is continuously decomposed by a autoradiolysis to $F_2$ and lower plutonium fluorides.

Autoradiolysis is expected to be more pronounced in hydrated materials because of the intervention of reactive species from radiolysis of water. On the other hand, the heat released in these materials may anneal part of the damage.

## 14.5 Radiation of TBP Solutions

Tributyl phosphate (TBP) has been used in the separation of thorium, uranium, and plutonium from nitrate solutions since the Manhattan Project of World War Ⅱ. Subsequently, it has come to be used world-wide in a process called PUREX (Plutonium Uranium Refining by Extraction), usually from 1 to 3 M aqueous nitric acid. The metal-loaded solvent is then stripped and recycled. However, its recycle potential is limited by the radiolytic degradation of TBP and its diluents. The accumulation of degradation products results in decreased performance including poor phase separation, decreased mass-transfer coefficients, and poor fission-product/actinide separation factors.

It has long been recognized that the major products of TBP radiolysis are hydrogen, methane, and dibutylphosphoric acid (HDBP), with monobutylphosphoric acid (H2MBP) and phosphoric acid produced in lesser amounts. While HDBP is a major product with consequences for solvent performance, its reported yield varies widely, and comparisons are difficult due to the differing experimental conditions used in the many studies that have been performed. Most investigations have focused on steady-state radiolysis using low LET isotopic gamma ray or accelerator electron sources, followed by stable-product identification. Only a few studies have been done using high LET radiation. Those have reported little difference in the yield of HDBP for $\alpha$ versus $\gamma$ irradiation.

Additional products of TBP radiolysis in the presence of $HNO_3$ are nitrated phosphates, which also impede stripping efficiency. It was proposed that $^{\cdot}NO_3$ reacts with TBP by hydrogen atom abstraction, producing the TBP radical:

$$^{\cdot}NO_3 + TBP \longrightarrow (C_4H_9O)_2(^{\cdot}C_4H_8O)PO + HNO_3 \tag{14.13}$$

The TBP radical was then postulated to undergo reaction with additional $^{\cdot}NO_3$ to produce nitrated TBP:

$$^{\cdot}NO_3 + (C_4H_9O)_2(^{\cdot}C_4H_8O)PO \longrightarrow (C_4H_9O)_2(OC_4H_8NO_3)PO \tag{14.14}$$

Nitrated phosphates ($RONO_2$ and $RNO_2$) have been identified in infra-red and mass spectrometric analyses of post-irradiation 30% TBP/dodecane/ $HNO_3$ solutions. The $^{\cdot}NO_3$ radical might be expected to add in the same way:

$$^{\cdot}NO_2 + (C_4H_9O)_2(^{\cdot}C_4H_8O)PO \longrightarrow (C_4H_9O)_2(OC_4H_8NO_2)PO \tag{14.15}$$

Methylated and hydroxylated phosphates, some of high molecular weight, have also been measured probably due to addition of hydroxyl, methyl, or another TBP radical.

Normal alkanes such as dodecane, Isopar L, and kerosene are the common organic diluents in nuclear solvent extraction. The radiolysis of alkanes is represented by:

$$CH_3(CH_2)_nCH_3 -\!\!\!\!\sim\!\!\sim\!\!\sim \longrightarrow e^-_{sol} + CH_3(CH_2)_nCH_3^{\cdot} + CH_3(CH_2)_nCH_2^{\cdot} + {}^{\cdot}CH_3 + H^{\cdot} + H_2 \tag{14.16}$$

The carbon-centreed radical products of Eq. (14.17) have a number of possible fates, including hydrogen atom abstraction from other species in solution or radical-radical addition to create higher molecular-weight products:

$$CH_3(CH_2)_nCH_2^{\cdot} + CH_3(CH_2)_nCH_2^{\cdot} \longrightarrow {}^{\cdot}CH_3(CH_2)_{2n+2}CH_3 \tag{14.17}$$

As the absorbed dose increases, this may result in changing physical characteristics of the solvent, including phase-disengagement time, density, and viscosity. Alkane radicals may also add to solute radicals to create alkylated solutes. Finally, two radicals may undergo disproportionation, to produce paraffinic and olefinic products:

$$CH_3(CH_2)_nCH_2^{\cdot} + CH_3(CH_2)_nCH_2^{\cdot} \longrightarrow CH_3(CH_2)_nCH_3 + CH_3(CH_2)_{n-1}CH = CH_2 \tag{14.18}$$

A common fate of the radical cation product of Eq. (14.19) ($CH_3(CH_2)_nCH_3^{\cdot+}$) in aqueous solution involves fragmentation via C-H bond cleavage to produce a carbon-centreed radical and a proton:

$$CH_3(CH_2)_nCH_3^{\cdot+} \longrightarrow CH_3(CH_2)_{n-1}({}^{\cdot}CH)CH_3 + H^+ \tag{14.19}$$

The radical cation product of Eq. (14.20) can also undergo charge transfer with another species, RH, that may be another solvent molecule resulting in no net effect, or an organic solute. An example of this is shown in Eq. (14.21):

$$CH_3(CH_2)_nCH_3^{\cdot+} + RH \longrightarrow CH_3(CH_2)_nCH_3 + RH^{\cdot+} \tag{14.20}$$

Finally, in these aerated solutions, neutral carbon-centreed radicals add $O_2$ to produce peroxyl radicals:

$$R^{\cdot} + O_2 \longrightarrow ROO^{\cdot} \tag{14.21}$$

Peroxyl radicals undergo radical-radical addition to form tetroxides, which then decompose to produce aldehydes, ketones, and alcohols from the original alkane. They are therefore important intermediates in the oxidative mineralization of organic compounds by radiolysis.

### 词汇

| | |
|---|---|
| ampoule | 安瓿(装针剂的小玻璃瓶) |
| radon | 氡 |
| delocalised charge | 离域电荷 |
| electromagnetic radiation | 电磁辐射 |
| ionization | 电离 |
| excitation | 激发 |
| annihilation process | 湮灭过程 |
| bremsstrahlung | 韧致辐射 |
| linear energy transfer | 线性能量传输 |
| diffuse | 扩散 |
| ozone | 臭氧 |
| nitrous oxide | 一氧化二氮 |
| peroxide | 过氧化物 |
| degradation product | 降解产物 |
| emanation power | 射气功率 |

| | |
|---|---|
| ephemeral isotope | 短暂核素 |
| adsorb | 吸附 |
| aerosol | 气溶胶 |
| radiological hazard | 放射性危害 |
| inhal | 吸入 |
| retain | 保留 |
| broncho-respiratory tract | 支气管呼吸道 |
| contamination | 污染 |
| aqueous solution | 水溶液 |
| dilute solution | 稀溶液 |
| free radical | 自由基 |
| hydrated electron | 水合电子 |
| fast mutual reaction | 快速相互反应 |
| reducing agent | 还原剂 |
| pleochroic halos | 多色光圈 |
| metamictization | 蜕晶质化 |
| autoradiolysis | 自辐射分解 |
| intervention | 干涉 |
| anneal | 退火 |
| dibutylphosphoric acid | 二丁基磷酸 |
| monobutylphosphoric acid | 一丁基磷酸 |
| mass spectrometry | 质谱 |

## 注释

1. Among the radiation effects observed were the fuorescence induced in different salts and the changes in their crystallographic form resulting in color changes from delocalised charges in the crystals.

在观察到的辐射效应中,有在不同的盐中辐照诱发的荧光,以及它们的晶体形式的变化,导致晶体中离域电荷的颜色变化。

2. In radiation chemistry, the *G* value is used as a measure of the chemical yield.

在辐射化学中,*G* 值被用来衡量化学产额。

3. The emanation power of the medium is the fraction of the formed radon atoms that is released into the surroundings.

介质中的射气功率是释放到周围环境中所形成的氡原子的分数。

4. For particles with high LET, such as ex particles and fission fragments, the density of primary species is very high and mutual reactions are favored. For fast electrons on the other hand, more free radicals diffuse into the bulk of the solutions.

对于 LET 高的粒子,如前粒子和裂变碎片,初级物种的密度非常高,有利于相互反应。另一方面,对于快速电子,更多的自由基将会扩散到大部分溶液中。

5. Radiation effects of solid compounds were first observed in the coloration of glassware in contact with radioelements, and in pleochroic halos in mica irradiated by the inclusion of radioactive substances.

固体化合物的辐射效应可以在与放射性元素接触的玻璃器皿的着色中观察到,也可以在放射性物质包裹物照射的云母中的多色光圈中观察到。

6. Autoradiolysis is expected to be more pronounced in hydrated materials because of the intervention of reactive species from radiolysis of water. On the other hand, the heat released in these materials may anneal part of the damage.

由于自辐射分解产生的反应物种的干预,预计自身放射性分解在水合材料中更为明显。另一方面,这些材料中释放的热量可能会使部分损伤退火。

7. The accumulation of degradation products results in decreased performance including poor phase separation, decreased mass-transfer coefficients, and poor fission-product/actinide separation factors.

降解产物的积累导致性能下降,包括较差的相分离、降低的传质系数和较差的裂变产物/锕系元素分离因子。

# Chapter 15 Spent Nuclear Fuel Reprocessing

## 15.1 Management of Spent Fuel

The fissioning of $^{235}U$ produces more neutrons than are needed to continue a chain reaction. Some of them combine with $^{238}U$ nuclei, which far outnumber $^{235}U$ nuclei in the reactor fuel. When $^{238}U$ captures a neutron, it is changed into $^{239}U$. The $^{239}U$ then decays into $^{239}Np$, which decays into $^{239}Pu$. This same process forms $^{239}Pu$ in a breeder reactor. Slow neutrons can fission $^{239}Pu$, as well as $^{235}U$. Some of the newly formed $^{239}Pu$ is thus fissioned during the fissioning of $^{235}U$. The rest of the $^{239}Pu$ remains in the fuel assemblies.

The fissioning of $^{235}U$ also produces many other radioactive isotopes, such as strontium 90, cesium 137, and barium 140. Fuel wastes can remain dangerously radioactive for hundreds of years because of the strontium and cesium isotopes. After that time, enough of the strontium and cesium will have decayed into stable isotopes so that they no longer present a severe problem. However, the plutonium and other artificially created elements from $^{238}U$ transmutation remain radioactive for thousands of years. Even in small amounts, plutonium can cause cancer or genetic damage in human beings. Larger amounts can cause radiation sickness and death. Safe disposal of these wastes is one of the most difficult problems involved in nuclear power.

Nowadays there are two main nuclear fuel cycle strategies: the once-through cycle (direct disposal or open cycle) and the twice-through cycle (recycling or partially closed cycle). The once-through cycle considers spent nuclear fuel (SNF) to be high-level waste (HLW) and, consequently, it is directly disposed of in a storage facility without being put through to any chemical processes, where it will be safely stored for millions of years until its radiotoxicity reaches natural uranium levels or another safe reference level. The twice-through cycle considers SNF to be an energy source due to its composition: approximately 96%-97% of its components are recyclable materials, 94%-96% of which is uranium (1% approx. of $^{235}U$) and 1%-1.5% is plutonium. Thus, in order to exploit its energy potential, SNF has to be put through a series of chemical processes known as reprocessing. Therefore, on the twice-through cycle strategy, SNF is reprocessed in order to extract the uranium and plutonium, which can either be recycled as fresh nuclear fuel for its use in a nuclear reactor that is adapted to this type of fuel or sold as raw material.

The once-through cycle comprises two main back end stages: interim storage and final disposal. In between those stages, there is also an encapsulation step and transportation. First, SNF is extracted from the reactor and cooled inside the reactor pools for a period of at least 5 to 10 a, though in some cases it was considered 3 a. This stage is mandatory for every nuclear fuel cycle strategy. After this period of initial cooling, nuclear fuel can be transferred into a dry interim storage, which can be either at the Nuclear Power Plant (NPP) site or at a centralized location that stores SNF from more than one NPP. Finally, after a minimum period of 50 to 100 a of interim storage, SNF must be transferred to a final disposal facility. Currently, the preferred option is a deep geological repository (DGR), which is an underground emplacement in stable geological formations.

The twice-through cycle is a little more complicated than the once-through cycle and, thus, involves more stages both in the back end and in the front end. After the initial cooling at the reactor pools, SNF is transported to the reprocessing facility, where the U and the Pu are separated from the minor actinides (MA) and fission products (FP), by means of the PUREX (Plutonium and Uranium Redox Extraction) process, which is currently the only commercially available technology. Once they have been separated, MAs and FPs are vitrified in a glass matrix and stored as HLW. Eventually, they will be transferred into a final disposal facility, most probably a DGR. However, since SNF is composed of more than 96% of U and Pu, the final HLW volume is reduced about 80%, its radiotoxicity decreases about 90% and its decay heat is also reduced compared to the once-through cycle. Thus, the cost of a DGR for vitrified waste would be about 25% of the cost of a DGR for SNF. Meanwhile, the extracted plutonium can be recycled into MOX fuel after being transferred to a MOX fuel fabrication plant. This type of fuel can be used in about 10% of the world's reactors, but currently the number of Pu recycles is limited to two or three, due to technological limitations. Thus, irradiated MOX has to be stored until its final disposition or until advanced reprocessing technologies are available. Additionally, uranium can also be reused after re-enrichment and fabrication at a dedicated plant. However, recovered uranium is more radioactive than natural uranium. Thus, the cost of the processes of enrichment and fuel fabrication are higher for recovered uranium than for natural uranium, because it requires dedicated facilities to avoid natural uranium contamination. Hence, this practice is not very common at this time, for this reason, reprocessed uranium is usually stored.

## 15.2 Spent Fuel Reprocessing

Nuclear fuel reprocessing is the chemical separation of fission products and unused uranium from spent nuclear fuel. The advantages of reprocessing compared to the direct disposal option come from the reduction of natural uranium requirements and the decrease in volume, radiotoxicity and decay heat of the final HLW, as long as irradiated MOX is treated separately. Consequently, the final repository volume is reduced, which also decreases its cost. Furthermore, HLW compaction and vitrification facilitates the handling of the final waste. However, there are some disadvantages associated with reprocessing due to the complexity of the nuclear fuel cycle, which creates more stages and fuel transportations. Moreover, reprocessing requires the extraction of the nuclear fuel, which increases the exposure risk, as well as LILW (low and intermediate lived waste) volumes. In addition, the separation of pure plutonium increases the proliferation risk. However, there are some options to solve this issue such as the recycling of Pu in MOX fuel, which reduces the Pu inventory, but also some advanced reprocessing techniques that jointly extract Pu and U, avoiding pure Pu separation. Evolution of nuclear fuel cycle costs is shown in Figure 15. 1.

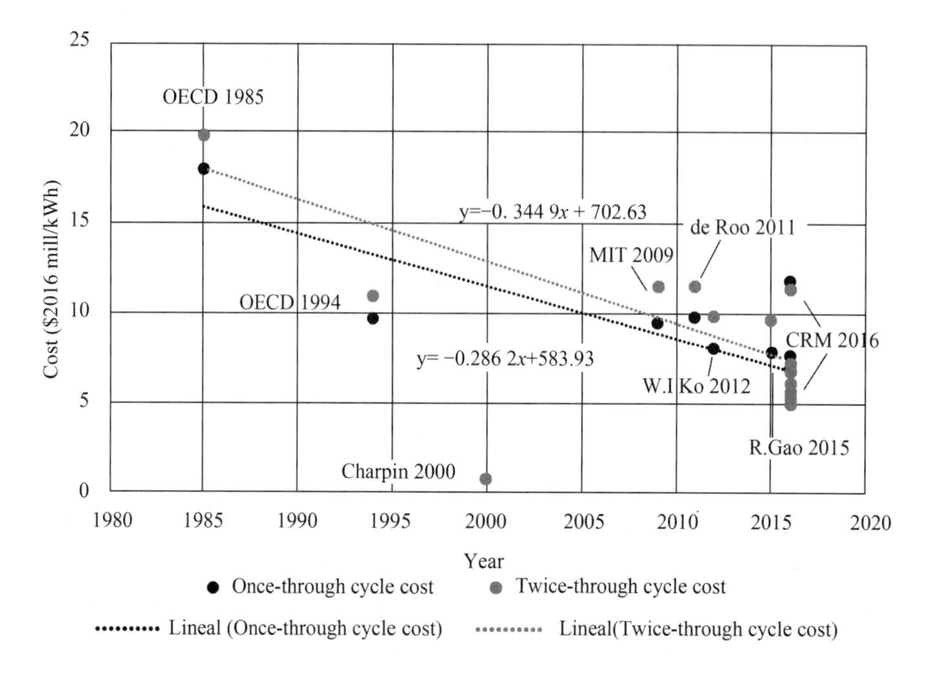


**Figure 15. 1　Evolution of nuclear fuel cycle costs**

Originally, reprocessing was used solely to extract plutonium for producing nuclear weapons. With commercialization of nuclear power, the reprocessed plutonium was recycled back into MOX nuclear fuel for thermal reactors. The reprocessed uranium, also known as the spent fuel material, can in principle also be re-used as fuel, but that is only economical when uranium supply is low and prices are high. A breeder reactor is not restricted to using recycled plutonium and uranium. It can employ all the actinides, closing the nuclear fuel cycle and potentially multiplying the energy extracted from natural uranium by about 60 times. Reprocessing must be highly controlled and carefully executed in advanced facilities by highly specialized personnel. Fuel bundles which arrive at the sites from nuclear power plants (after having cooled down for several years) are completely dissolved in chemical baths, which could pose contamination risks if not properly managed. Relatively high cost is associated with spent fuel reprocessing compared to the once-through fuel cycle, but fuel use can be increased and waste volumes decreased.

Most commercial and developing processes for actinide separation from spent nuclear fuel are presently solvent extraction processes. An organic solvent containing an extractant ligand is contacted with an aqueous solution of spent nuclear fuel, usually dissolved in nitric acid. The organic extractant is chosen or designed to have the chemical ability to selectively partition the desired elements from the aqueous phase into the organic solvent. An ideal extractant should be efficient, selective, hydrolytically and radiolytically stable, soluble, give fast reaction kinetics and be easily synthesized.

Due to the large number of elements present in spent nuclear fuel, many varied solvent extraction processes have been developed to selectively extract different groups of elements. These processes may be classed into three groups corresponding to the three main sequential solvent extraction processes considered and utilised for treatment of spent nuclear fuel. The initial step is to recover the U and perhaps also Pu for recycling as nuclear fuel. This process is presently performed on an industrial scale using solutions of spent nuclear fuel dissolved in concentrated nitric acid. Therefore, the subsequent solvent extraction processes, involving removal of both MA and Ln together from the remaining fission products in the raffinate after U and Pu removal, must also be performed with nitric acid as the aqueous phase. The final step then involves separation of MA from Ln, allowing the isolated MA to be transmutated into less hazardous materials.

## 15.3 PUREX Reprocessing Process

PUREX (Plutonium and Uranium Recovery by Extraction) is used for almost all fuel reprocessing today. Irradiated $UO_2$ fuel is dissolved in $HNO_3$ with the uranium being oxidized to $UO_2(NO_3)_2$ and the plutonium oxidized to $Pu(NO_3)_4$. A solution of tributyl phosphate(TBP) in a high-boiling hydrocarbon, such as n-dodecane, is used to selectively extract the hexavalent $UO_2(NO_3)_2$ and the tetravalent $Pu(NO_3)_4$ from the other actinides and fission products in the aqueous phase. The overall reactions are

$$UO_2^{2+}(aq) + 2NO_3^-(aq) + 2TBP(org) \longrightarrow UO_2(NO_3)_2 \cdot 2TBP(org) \tag{15.1}$$

or

$$Pu^{4+}(aq) + 4NO_3^-(aq) + 2TBP(org) \longrightarrow Pu(NO_3)_4(TBP)_2(org) \tag{15.2}$$

These equilibria can be shifted to the right (i. e., improved extraction) by increasing the TBP concentration in the organic phase or increasing the $[NO_3^-](aq)$. In a second step, the TBP solution is treated with a dilute nitric acid solution of a reducing agent, such as ferrous sulfamate or U(Ⅳ), which reduces the plutonium to a trivalent state but leaves the uranium in a hexavalent state. Plutonium will then transfer to the aqueous phase, leaving uranium in the organic phase. The uranium is stripped from the organic phase.

The only fission fragments that extract during the PUREX process are Zr, Ru, Nb, and Tc, with the most troublesome being Zr and Ru. Zr forms a number of complex species with the most important being $[Zr(NO_3)_4 \cdot 2TBP]$. The formation of this complex is inhibited by the addition of $F^-$ whereby

$$Zr(NO_3)_4 \cdot 2TBP + 6F^- \rightleftharpoons ZrF_6^{2-} + 4NO_3^- + 2TBP \tag{15.3}$$

An overall schematic view of the PUREX process is shown in Figure 15.2. The first step is to prepare the irradiated fuel for dissolution by mechanically chopping it into small pieces (1–5 cm). This opening of the cladding causes the release of 10% of the Kr and Xe fission products as well as some $^3H$ and volatile fission products. These off-gases are combined with those from the dissolution step. In the dissolution step, the fuel pieces are dissolved in near boiling 10 M $HNO_3$. This step, which takes a few hours, dissolves the uranium, plutonium, and fission products, leaving the cladding to be recovered. The Kr and Xe are recovered from the off-gas of steam, air, and $NO_x$. The chemical reactions for the dissolution of uranium involve processes like

$$3UO_2 + 8HNO_3 \longrightarrow 3UO_2(NO_3)_2 + 2NO + 4H_2O \tag{15.4}$$

and

$$UO_2 + 4HNO_3 \longrightarrow UO_2(NO_3)_2 + 2NO_2 + 2H_2O \tag{15.5}$$

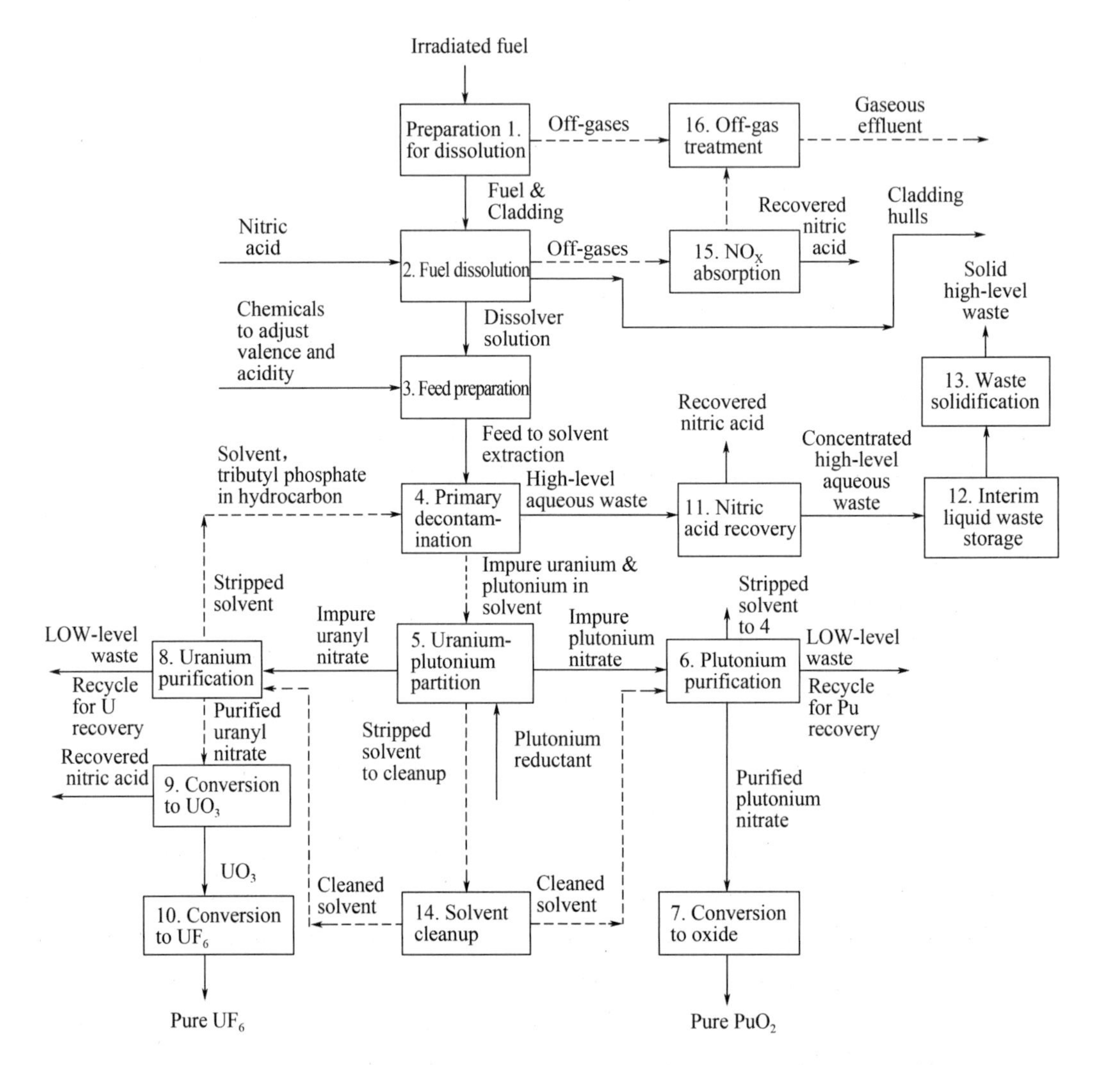


**Figure 15.2 Schematic diagram of the PUREX process(From Benedict et al., 1981.)**

The plutonium is oxidized to Pu(Ⅳ) and Pu(Ⅵ), while the neptunium ends up in the pentavalent or hexavalent states. Small amounts of plutonium and fission products may not dissolve, and they can be leached with acid solutions containing the oxidant $Ce^{4+}$.

The dissolver solution is treated with chemicals to adjust the acidity, valence, and concentrations of the species involved. The $HNO_3$ concentrations are about 2-3 M, the $UO_2(NO_3)_2$ concentrations are about 1-2 M, and the Pu is stabilized as Pu(Ⅳ) using $N_2O_4$ or hydroxylamine. In these and subsequent manipulations of these solutions, attention must be given to criticality control. This is done by regulating the solution geometry, the concentrations of fissile materials, and by the addition of neutron absorbers such as Gd.

The primary separation of plutonium and uranium from the fission products involves a solvent extraction with 30 vol. % TBP at room temperature. The activity levels in this separation are quite high and the aqueous waste, which contains 99% of the fission products, is a high-level waste. Am and Cm are not extracted and Np is partially extracted. Because of the high radiation levels, there are radiolysis problems with TBP, leading to solvent degradation. Primary products of the radiolysis of TBP are the dibutyl and monobutyl phosphoric acids along with phosphoric acid. These degradation products are removed in the solvent purification steps.

Following decontamination of the uranium/plutonium from the fission products, the plutonium is separated from the uranium. This is done by reducing the Pu (Ⅳ) to nonextractable Pu(Ⅲ), leaving uranium in the hexavalent state. In the older Purex plants, this was done using $Fe^{2+}$ while the newer plants add $U^{4+}$. The plutonium thus ends up in an aqueous phase while the uranium remains in the organic phase.

Uranium is back-extracted (and thus removed from the organic phase) with 0.01 M $HNO_3$. It is purified by a series of solvent extraction cycles until the Pu/U ratio is $<10^{-8}$ and the total $\beta$ and $\gamma$ activity is less than twice that of aged natural uranium.

## 词汇

| | |
|---|---|
| nuclear fuel reprocessing | 核燃料后处理 |
| strontium | 锶 |
| cesium | 铯 |
| barium | 钡 |
| transmutation | 嬗变 |
| waste management | 废物处理 |
| nuclear fuel cycle | 核燃料循环 |
| nuclear power plant | 核电站 |
| spent nuclear fuel | 乏燃料 |
| once-through cycle | 一次通过循环 |
| twice-through cycle | 两次通过循环 |
| high-level waste | 高放废物 |
| radiotoxicity | 放射毒性 |
| component | 组成 |
| extract | 萃取 |
| interim storage | 临时贮存 |
| deep geological repository | 深地质处置库 |
| Plutonium and Uranium Redox Extraction | 钚-铀共同氧化还原萃取(普雷克斯流程) |
| mixed oxide | 混合氧化物 |

| | |
|---|---|
| fission product | 裂变产物 |
| vitrification | 玻璃固化 |
| low and intermediate lived waste | 中低放废物 |
| thermal reactor | 热反应堆 |
| fuel bundle | 燃料棒 |
| solvent extraction process | 溶剂萃取流程 |
| sequential | 连续的 |
| hazardous | 有毒的 |
| tributyl phosphate | 磷酸三丁酯 |
| reductant | 还原剂 |
| acetohydroxamic acid | 乙酰氧肟酸 |
| extractability | 萃取率 |
| proliferation resistance | 防扩散 |
| product stream | 产品流 |
| raffinate | 萃余液 |
| alpha activity | α 放射性 |
| degradation | 降解 |
| specific activity | 比活度 |
| hydrochloric acid | 盐酸 |
| reextraction | 反萃 |

## 注释

1. Nowadays there are two main nuclear fuel cycle strategies: the once-through cycle (direct disposal or open cycle) and the twice-through cycle (recycling or partially closed cycle).

目前有两种主要的核燃料循环战略:一次循环(直接处置或开放式循环)和两次循环(再循环或部分封闭式循环)。

2. The advantages of reprocessing compared to the direct disposal option come from the reduction of natural uranium requirements and the decrease in volume, radiotoxicity and decay heat of the final HLW, as long as irradiated MOX is treated separately.

与直接处置方案相比,后处理的优势在于减少了对天然铀的需求,并降低了最终高放废物的体积、放射毒性和衰变热,只要对经过辐照的混合氧化物进行单独处理即可。

3. An ideal extractant should be efficient, selective, hydrolytically and radiolytically stable, soluble, give fast reaction kinetics and be easily synthesized.

理想的萃取剂应具有高效性、选择性、水解稳定性和放射性稳定性、可溶性、反应动力学快以及易于合成等特点。

4. SNF is composed of more than 96% of U and Pu, the final HLW volume is reduced about 80%, its radiotoxicity decreases about 90% and its decay heat is also reduced compared to the

once-through cycle.

乏燃料由 96% 以上的铀和钚组成,与一次通过循环相比,最终高放废物体积减少了约 80%,其放射性毒性降低了约 90%,衰变热也减少了。

5. A solution of TBP in a high-boiling hydrocarbon, such as n-dodecane, is used to selectively extract the hexavalent $UO_2(NO_3)_2$ and the tetravalent $Pu(NO_3)_4$ from the other actinides and fission products in the aqueous phase.

TBP 在高沸点碳氢化合物,如正十二烷中的溶液,用于从水相中的其他锕系元素和裂变产物中选择性地萃取六价 $UO_2(NO_3)_2$ 和四价 $Pu(NO_3)_4$。

6. The first step is to prepare the irradiated fuel for dissolution by mechanically chopping it into small pieces (1-5 cm).

第一步是用机械方法将辐照燃料切成小块(1~5 厘米),为溶解做好准备。

7. The plutonium is oxidized to Pu(Ⅳ) and Pu(Ⅵ), while the neptunium ends up in the pentavalent or hexavalent states. Small amounts of plutonium and fission products may not dissolve, and they can be leached with acid solutions containing the oxidant $Ce^{4+}$.

钚被氧化成钚(Ⅳ)和钚(Ⅵ),而镎最终以五价或六价状态存在。少量的钚和裂变产物可能不会溶解,它们可以被含有氧化剂 $Ce^{4+}$ 的酸溶液浸出。

# Chapter 16 Modification of PUREX Reprocessing Process

## 16. 1 UREX Process

The PUREX process can be modified to make a UREX (URanium EXtraction) process which could be used to save space inside high level nuclear waste disposal sites, such as the Yucca Mountain nuclear waste repository, by removing the uranium which makes up the vast majority of the mass and volume of used fuel and recycling it as reprocessed uranium. The UREX process is a PUREX process which has been modified to prevent the plutonium from being extracted. This can be done by adding a plutonium reductant before the first metal extraction step. In the UREX process, about 99. 9% of the uranium and >95% of technetium are separated from the other fission products and actinides. The key is the addition of acetohydroxamic acid (AHA) to the extraction and scrub sections of the process. The addition of AHA greatly diminishes the extractability of plutonium and neptunium, providing somewhat greater proliferation resistance than with the plutonium extraction stage of the PUREX process. As with PUREX, 30% TBP in kerosene is used to extract U(Ⅵ) and Pu(Ⅳ) from dissolved spent nuclear fuel. Then, a subsequent scrubbing of the loaded organic phase with 0. 5 M AHA in 0. 3 M nitric acid is used to return the Pu(Ⅳ) to the aqueous phase since the AHA forms a strong complex with Pu(Ⅳ) but does not extract U(Ⅵ). Substantial amounts of Tc are also extracted with the U(Ⅵ) into the organic phase, so the Tc is stripped using 6 M nitric acid before final conversion of the U(Ⅵ) to an oxide that can be re-used as fuel.

UREX and variants of UREX processes are used for reprocessing of the spent nuclear fuels. The UREX process consists of nine versions from which seven product streams are obtained. Uranium and technetium are recovered as separate products with the yield >99% in the first stage of the UREX process. Cesium and strontium are recovered from the UREX waste. Plutonium and neptunium are separated using NPEX process with high level of impurity. The NPEX raffinate is treated in the TRUEX process from which minor actinides along with rare-earth elements are separated. Purification of minor actinides from the rare-earth elements is carried out in the Cyanex 301 process.

The UREX flowsheet consists of three primary separations processes. The first process is a modification of the industrially used continuous solvent extraction process using tributyl phosphate. Largest-mass component, uranium, is removed, whereas plutonium and neptunium (possibly with some uranium) are collected together. The second process, reverse TALSPEAK, is a continuous solvent extraction, using di-2-ethyl-hexyl phosphoric acid, to separate and recover the minor actinides, americium and curium. The third process, also a solvent extraction, is designed to partition the fission product wastes to separate the high-heat generators, cesium and strontium, for managed storage.

Cesium is a major product of uranium fission, which is the most commonly existed radionuclide in radioactive wastes. Various technologies have been applied to separate radioactive cesium from radioactive wastes, such as chemical precipitation, solvent extraction, membrane separation and adsorption. Crown ethers and calixarenes derivatives can selectively coordinate with cesium ions by ion-dipole interaction or cation-$\pi$ interaction, which are promising extractants for cesium ions due to their promising coordinating structure.

## 16.2 TRUEX Process

Adding a second extraction agent, octyl(phenyl)-N, N-dibutyl carbamoylmethyl phosphine oxide (CMPO) in combination with tributyl phosphate(TBP), the PUREX process can be turned into the TRUEX (TRansUranic EXtraction) process. TRUEX was invented in the US by Argonne National Laboratory and is designed to remove the transuranic metals (Am/Cm) from waste. The idea is that by lowering the alpha activity of the waste, the majority of the waste can then be disposed of with greater ease. In common with PUREX this process operates by a solvation mechanism.

CMPO can effectively extract all trivalent and above actinides from a high concentration of 3 to 6 mol/L nitric acid, but its selectivity for trivalent actinides and lanthanide is low, so that lanthanides are co-extracted into the organic phase and eluted into nitric acid with americium and curium. At the same time, zirconium and molybdenum can co-extract with actinium and lanthanum in the extraction cycle, so precautions must be taken.

The extraction agent commonly used is a solution consisting of dodecane, CMPO and tributyl phosphate(TBP). TBP prevents the formation of third-phase, making the process more adaptable to acidity and phase stability, and reducing the hydrolysis and radiative degradation of CMPO. These extractants have high partition coefficient for extracting trivalent, tetravalent and hexavalent actinides from moderately acidic solutions and have good selectivity for most of the fission elements. The fact that plutonium (Ⅳ), uranium (Ⅵ), and americium (Ⅲ) have stable distribution coefficients at 1 to 6 mol/L nitric acid, which enables the extraction of actinides from nuclear liquid waste without or slightly adjusting the acidity of the feed. This is an important feature that distinguishes TRUEX from other processes.

The TRUEX process was validated with real high level liquid waste at the Idaho Chemical Treatment Plant. The acidity of high level liquid waste was 1.72 mol/L. The extraction agent consisted of 0.2 mol/L CMPO, 1.4 mol/L TBP, Isopar L (Iso-alkanes diluent). The experiment was carried out in a 24-stage (8-stage extraction, 5-stage washing, 6-stage back extraction, 5-stage extractant regeneration) centrifugal extractor. The total removal rate of actinide elements in the material liquid reached 99.97%. The specific activity of $1.75\times10^4$ Bq/g dropped to 4.44 Bq/g, much lower than the 40 Bq/g standard for non-alpha waste.

In addition, Los Alamos National Laboratory developed the TRUEX process for separating actinides from chloride aqueous solutions by dissolving waste chloride solution containing plutonium and americium in hydrochloric acid as feed, and using 0. 5 mol/L CMPO solution of tetrachloroethylene as actinide extractant.

Shortcomings of TRUEX process: (1) In order to prevent the formation of third phase, TBP is added to CMPO as a modifier, which makes the extraction system complicated. (2) Cross-contamination exists among the logistics of various products. (3) Reextraction of neptunium and plutonium with HF aggravates the corrosion equipment for reprocessing. (4) Acid impurities resulting from hydrolysis and radiolysis of CMPO are detrimental to americium reextraction.

## 16. 3 COEX™ Process

The COEX™ or COEXtraction process has been developed by French workers, and is designed to bleed some of the U (and possibly Np) into the Pu product to eliminate the pure Pu stream. The development of this new reprocessing and recycling COEX™ process was aimed at:

(1) Further enhancing proliferation resistance;

(2) Maintaining a high level of process performance;

(3) Minimizing both investment (capital) and operational costs;

(4) Taking full advantage of present industrial experience;

(5) Keeping open possible evolutions to take into account new types of reactors or future changes in management strategies of the transuranic elements.

The separation flowsheets used in the COEX™ process are based on the expertise accruing from the design, and deployment of the PUREX process, but equally from operational feedback with that process. The design for the extraction cycles has extensively relied on a simulation tool, validated by experimental investigations carried out in the workshops at the La Hague Plants, and equally by comparisons with findings from industrial operations. Consequently, the COEX™ process is basically an evolution of the PUREX process, by modifying it to produce a U+Pu mixture (U/Pu >20%), rather than pure plutonium: no pure plutonium is separated at any point of the process. The advantage of so doing is to curb proliferation risks and to produce a perfectly homogeneous mixed oxide, for the purposes of MOX fuel fabrication, affording enhanced performance.

The proposed COEX™ flowsheet involves an initial U/Pu codecontamination step that is fairly similar to the first purification cycle. The codecontamination step is unchanged: this covers the extraction flowsheet, including measures to allow removal of technetium. The plutonium stripping part is modified by including conditions that results in uranium scrubbing into the Pu strip product. This function is slightly tweaked from uranium scrub to neptunium scrub, to allow some uranium to remain in the plutonium stream. The function further allows neptunium to be extracted from the plutonium stream, directing the neptunium thus extracted to the cycle's uranium product stream, thanks to higher distribution coefficient of Np(Ⅳ) compared to U(Ⅳ).

To ensure the plutonium becomes extractable by the solvent phase again, an adjustment is made to the solution yielded by the first cycle, by raising the nitric acid concentration, concomitant with reoxidation of plutonium to oxidation state Ⅳ, uranium being adjusted to oxidation state Ⅵ. The stream then undergoes treatment in the form of a U-Pu cycle, only involving plutonium extraction, and back-extraction (stripping) functions. Reductive plutonium stripping is effected using hydroxylamine nitrate.

Topping up with uranium is provided for, at the head end of this operation, to adjust the Pu/U ratio in the production stream. Finally, the stripped solvent is recycled in the plutonium stripping operation carried out upstream, in the first cycle. This measure makes it possible to tolerate some loss of plutonium to the stripped solvent, and thus limit the number of stages used for the plutonium stripping operation (hence the term "mini-cycle" for this complementary purification cycle).

A uranium purification cycle, identical to the cycles implemented at the La Hague plant, is further provided for — on the one hand, to complete of the uranium stream, with respect to β and γ emitters, and, on the other hand, to ensure removal of neptunium.

The behavior of Tc in the COEX™ process is basically the same as in PUREX flowsheet. More than 95% of Tc from the dissolution solution is recovered in raffinate of the complementary extraction, so in an effluent presenting a relative low activity (the bulk of the other fission products is eliminated during the first extraction-scrubbing steps). The other 5% of Tc is recovered in the raffinate of the second U/Pu purification cycle ("mini-cycle").

The bulk of the Np is recovered in the raffinate of the uranium purification cycle. Variants for the flowsheet have, at the same time, already been suggested that allow for evolutions in the near future, that would cater to neptunium recycling. A very simple modification of Pu stripping flowsheet in first extraction cycle (suppression of the Np scrub and use of HAN as reducing agent) allows neptunium to follow the plutonium-uranium stream. Then in Pu-U purification cycle, with a minor flowsheet modification, neptunium will follow U-Pu stream up to concentration stage.

## 16.4　SANEX Process

Considerable efforts have been made recently to develop advanced separation methodologies in order to maximize fuel resources and reduce the impact of nuclear waste while providing a proliferation-resistant fuel cycle (i. e., no pure plutonium is isolated). This forms part of the "Partitioning and Transmutation" strategy, where it is proposed that all of the actinides in SNF, including the minor actinides, can be separated and recycled as nuclear fuel. Another option is to "burn" the separated actinides, which will also result in conversion to short-lived fission product nuclides but without nuclear energy production for public consumption. This provides the added benefit of converting most of the long-lived actinides in SNF to shorter-lived fission product nuclides compared to current spent fuel management options. As a result, the "Partitioning and Transmutation" strategy can significantly reduce the time it takes for SNF to decay to radioactivity levels of natural uranium and therefore the necessary design lifetime of any nuclear waste repository. One of the major separation challenges that need to be overcome for this strategy to be successful is the separation of americium and curium from the lanthanide fission products. This is because the high neutron absorption cross sections of some of the lanthanide ions present in SNF both decrease the flux in a reactor and create more activation products, thereby making transmutation a less attractive option if the lanthanides cannot be separated from the actinides. Achieving this separation is extremely difficult because of the chemical similarities between americium, curium, and the lanthanides, which all most commonly exist in the Ⅲ oxidation state in solution. Consequently, organic molecules that can selectively extract actinides, in particular $Am^{3+}$ and $Cm^{3+}$, over the $Ln^{3+}$ ions are of great interest, as is evident by the number of different ligand systems and processes that have been developed by various groups in the field of partitioning.

Selective ActiNide EXtraction. As part of the management of minor actinides, it has been proposed that the lanthanides and trivalent minor actinides should be removed from the PUREX raffinate by a process such as DIAMEX or TRUEX. To allow the actinides such as americium to be either reused in industrial sources or used as fuel, the lanthanides must be removed. The lanthanides have large neutron cross sections and hence they would poison a neutron driven nuclear reaction. To date the extraction system for the SANEX process has not been defined, but currently several different research groups are working towards a process. For instance, the French CEA is working on a bis-triazinyl pyridine (BTP) based process. Other systems such as the dithiophosphinic acids are being worked on by some other workers.

### 词汇

| | |
|---|---|
| extractability | 可萃取性 |
| proliferation resistance | 抗增殖性 |
| substantial | 大量的 |
| product stream | 产品流 |
| impurity | 杂质 |
| raffinate | 萃余液 |
| modification | 修改 |
| hexavalent | 六价的 |
| co-extracted | (被)共萃取 |
| radiative degradation | 辐解 |
| hydrolysis | 水解作用 |
| moderately acidic | 中等酸度的 |
| distribution coefficients | 分配系数 |
| modifier | 相改性剂 |
| reoxidation | 再氧化 |
| considerable | 相当大的 |
| proliferation-resistant | 抗增殖性 |
| partitioning and transmutation | 分离和嬗变 |

absorption 吸收、吸附
similarity 相似、类似

## 注释

1. The addition of AHA greatly diminishes the extractability of plutonium and neptunium, providing somewhat greater proliferation resistance than with the plutonium extraction stage of the PUREX process.

AHA 的加入大大降低了钚和镎的可提取性,与 PUREX 工艺的钚提取阶段相比,提供了更大的抗核扩散能力。

2. Crown ethers and calixarenes derivatives can selectively coordinate with cesium ions by ion-dipole interaction or cation-π interaction, which are promising extractants for cesium ions due to their promising coordinating structure.

冠醚和杯芳烃衍生物可以通过离子-偶极子相互作用或阳离子-π 相互作用与铯离子选择性配位,具有良好的配位结构,是很有前途的铯离子萃取剂。

3. TBP prevents the formation of third-phase, making the process more adaptable to acidity and phase stability, and reducing the hydrolysis and radiative degradation of CMPO.

TBP 阻止了第三相的形成,使工艺对酸度的适应性和相的稳定性更强,减少了 CMPO 的水解和辐射降解。

4. To ensure the plutonium becomes extractable by the solvent phase again, an adjustment is made to the solution yielded by the first cycle, by raising the nitric acid concentration, concomitant with reoxidation of plutonium to oxidation state Ⅳ, uranium being adjusted to oxidation state Ⅵ.

为了确保钚再次被溶剂相提取,对第一个循环产生的溶液进行调整,通过提高硝酸浓度,同时将钚再氧化到氧化态Ⅳ,将铀调整到氧化态Ⅵ。

5. Considerable efforts have been made recently to develop advanced separation methodologies in order to maximize fuel resources and reduce the impact of nuclear waste while providing a proliferation-resistant fuel cycle (i. e., no pure plutonium is isolated).

最近为发展先进的分离方法做出了相当大的努力,以便最大限度地利用燃料资源和减少核废料的影响,同时提供防止扩散的燃料循环(即不分离纯钚)。

# Chapter 17 Pyroprocessing

## 17.1 General Description

Pyroprocessing is currently considered an alternative reprocessing technology to the more commonly used aqueous processing technology that accomplishes separations by way of high-temperature electrorefining. It has yet to be implemented on a large scale, limited to date to laboratory-scale and engineering-scale experimentation and demonstration. Much of the current state of the art for pyroprocessing was developed during the Integral Fast Reactor (IFR) program, which was carried out at Argonne National Laboratory from about 1984 to 1995. With the shut down of Experimental Breeder Reactor-Ⅱ in 1995, the IFR program was converted into a spent fuel treatment program to safely treat the 25 MT of heavy metal from that reactor. Pyroprocessing utilizes molten salt electrolytes as the media rather than acidic aqueous solutions and organic solvents. These electrolytes are principally used to support electrochemical separations such as uranium electrorefining and electrolytic reduction of oxide fuel. The process includes vacuum furnaces that accomplish salt/metal separations and melt metal deposits into ingots for either waste disposal or fuel fabrication. Ceramic and metal waste streams are generated that immobilize fission products and, optionally, plutonium and minor actinides into high level waste forms. For eventual commercial implementation, it is expected that plutonium and minor actinides will be recycled and used for fast reactor fuel fabrication. While this technology has yet to reach the commercialization stage, it has been the subject of extensive, government-funded research and development worldwide in addition to the EBR-Ⅱ spent fuel treatment work in the U. S. For example, the Republic of Korea is currently pursuing a strategy of developing pyroprocessing technology for treatment of spent fuel from their commercial light water reactor to minimize volume of high-level waste and possibly extract fissile actinides for eventual fabrication of fast reactor fuel.

While PUREX and related aqueous reprocessing technology have superior maturity, pyroprocessing does have unique benefits that make it a credible alternative and in some cases a preferred alternative. This includes use of process liquids that are more stable than organics in the presence of high radiation fields, improved criticality safety due to the lack of neutron moderators in the process, and waste processing that is integrated with the separation flowsheet.

## 17.2 Process Technology

There are many variants of the pyroprocessing flowsheet, but the IFR scheme shown in Figure 17.1 can be used as a reference, as it contains all of the key unit operations.

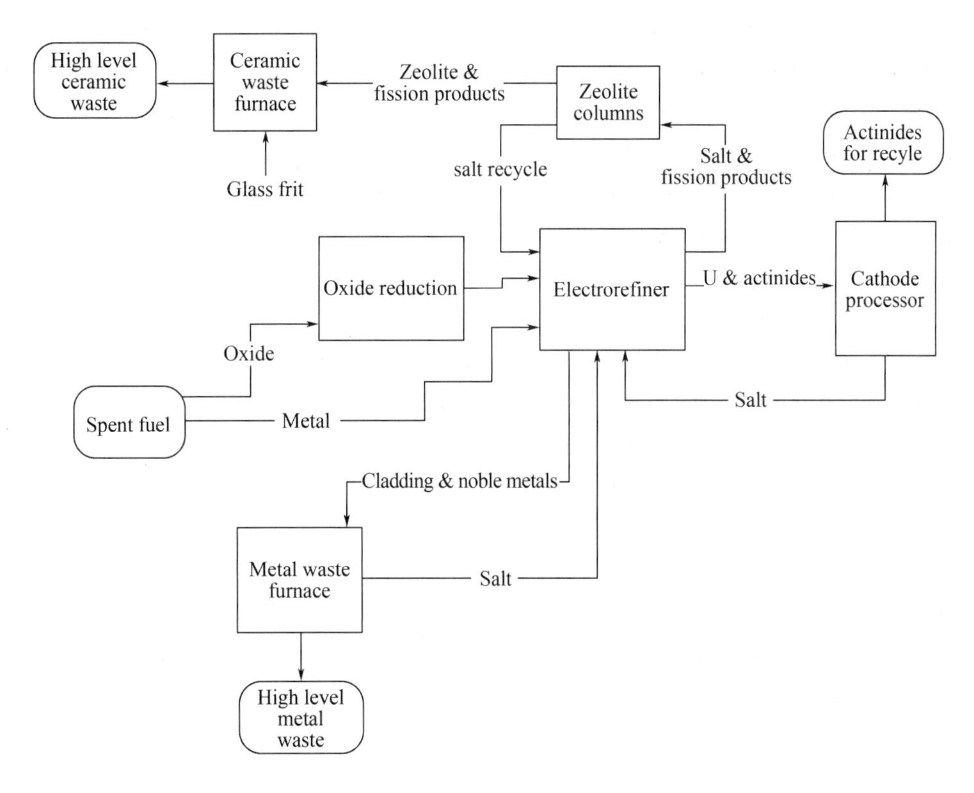


**Figure 17. 1 Fuel processing flowsheet for the integral fast reactor program**

The electrorefiner is at the centre of the flowsheet and is used to perform the primary separation of actinides from fission products. It contains a molten salt electrolyte typically LiCl-KCl-$UCl_3$ maintained at 450–500 ℃. The eutectic composition of LiCl-KCl (44. 2 wt% LiCl, 55. 8 wt% KCl) is maintained to keep the melting point at approximately 350 ℃. The $UCl_3$ content varies depending on desired operating conditions from about 0. 5 wt% to 10 wt%. It is used as a charge carrier for electrotransport through the electrolyte. After the spent fuel is chopped into segments, it is loaded into anode baskets, and the baskets are lowered into the electrorefiner. As current is passed between the anode and cathode, U metal is oxidized to $U^{3+}$ at the anode and reduced back to metallic form at the cathode. The deposit contains high purity uranium and is typically dendritic.

Transuranium (TRU) elements and active metal fission products are oxidized electrochemically or via reaction with uranium chloride in the salt and enter the electrolyte. Under normal conditions, Pu and minor actinides cannot deposit at the cathode, because their back-reaction with $UCl_3$ is thermodynamically spontaneous. However, co-deposition of U and TRU can be achieved via a combination of elevating the TRU to U ratio in the salt and utilizing a liquid cadmium cathode (LCC). In the molten cadmium phase, TRU elements have a very low activity coefficient compared to U. This allows for TRU metals to be present in quantities comparable to that of uranium in the cadmium. Alternative methods are currently being investigated to co-recover U and TRU without the need for an LCC.

Fission product elements segregate between the anode basket and the molten salt during the electrorefining process. Noble metals such as Tc, Ru, and Rh remain with the cladding hulls in the anode basket. Active metals that typically include Group Ⅰ and Ⅱ elements in addition to lanthanides are oxidized to chloride form and accumulate in the salt. If sodium metal is used as a bonding agent, as in the case of EBR-Ⅱ fuel, this sodium is oxidized to sodium chloride, which accumulates in the ER electrolyte.

Both metal fuel and oxide fuel can be treated via pyroprocessing. Treatment of metallic fuel is relatively straightforward due to the fact that it is already in a state compatible with the ER. Oxide fuel must first be converted to metallic form. This can be accomplished in an oxide reduction step. Various methods have been investigated for reducing spent oxide fuel. Early efforts were focused on chemical reduction via lithium. Similar to electrorefining, a molten salt is used for carrying out this reaction. In this case, it is LiCl saturated with lithium metal at 650 ℃. The lithium reduction reaction is as follows.

$$4Li + UO_2 \longrightarrow U + 2Li_2O \tag{15.1}$$

More recently, a similar process based on electrochemical reactions has been favored for development. It also uses a vessel containing molten LiCl at 650 ℃, but it contains lithium oxide in the salt rather than lithium metal. The reactions for the electrolytic process are shown below.

$$UO_2 + 4e^- \longrightarrow U + 2O^{2-} \tag{15.2}$$

$$2O^{2-} \longrightarrow O_2 + 4e^- \tag{15.3}$$

The generated oxygen bubbles out of the salt as a gas and can be sent to an off-gas treatment system to remove any entrained or volatile contaminants. The advantages of the electrolytic method based on the above reactions are that lithium oxide concentration in the molten salt can be kept low (about 1 wt%), and there is no need for a separate vessel to regenerate lithium metal from lithium oxide.

U or U/TRU product deposited on the cathode in the electrorefiner is transferred to a cathode processor, which is essentially a vacuum distillation furnace. The salt is separated from the metals and recycled to the electrorefiner. The purified metals can be fed into a process for fabricating metallic fuel for fast reactors. For the case of the EBR-Ⅱ spent fuel treatment process, the cathode processor operates at a temperature up to 1 200 ℃ and achieves pressures less than 1 torr.

After an electrorefining run, the anode basket contains the cladding hulls, undissolved actinides, inert fuel matrix material such as zirconium, adhering salt, and noble metal fission products such as Tc, Mo, Rh and Ru. All of this material is removed from the anode baskets and loaded into an inductively heated vacuum distillation furnace that is used to distill adhering salt and consolidate the metals into an ingot. The metal ingot becomes a waste form that has been tested and shown to be suitable for disposal in a high level waste repository.

Electrorefiner salt becomes progressively more contaminated with fission product chlorides as well as sodium chloride in the process of treating fuel. Once the contamination level has exceeded a pre-determined limit, the salt must be removed from the electrorefiner and either disposed or processed through a purification step and returned to the electrorefiner. The basis for that limit can be fission product decay heat, salt melting point, or criticality limits. Another potential limiting factor is contamination of the metallic actinide products recovered in the cathode processor. High concentrations of rare earth fission products in the salt, for example, have been shown to lead to high rare earth contamination levels in the actinide product. The process flowsheet shown in Figure 17. 1 includes zeolite ion exchange columns for achieving this salt purification. Zeolite-A has been shown to exhibit selectively for the fission product ions when in contact with molten chloride salt. Other alternatives that have been considered for treating the salt to remove fission products and other contaminants include selective precipitation, zone freeze refining, and adsorption by non-zeolitic materials.

The current baseline technology for dealing with the salt waste from electrorefining EBR-Ⅱ spent fuel is to non-selectively immobilize the salt into a ceramic waste form consisting of glass-bonded sodalite. In this process, salt is removed from the electrorefiner, sized via crushing and milling to a fine particulate, and adsorbed into zeolite-4A in a high temperature blending operation. A V-blender capable of heating and mixing particulate material to 500 ℃ is used for this absorption step. Prior to being loaded into the V-blender, it is necessary to dry the zeolite to less than 1 wt% water. This drying is used to maximize salt adsorption in the zeolite while minimizing evolution of water vapor in a high temperature, corrosive environment. Drying the zeolite should also minimize pores in the final ceramic waste form. Zeolite drying is accomplished via mechanically fluidizing the zeolite under vacuum at temperatures up to 550 ℃. Heating the zeolite-4A to temperatures of 600 ℃ or higher has been determined to cause structural damage that inhibits its ability to absorb salt. Final consolidation into the ceramic waste form occurs after borosilicate glass binder has been mixed with the salt-loaded zeolite, loaded into a steel canister, and heated to a maximum temperature of 915–950 ℃. During the process of consolidation, the zeolite-A phase converts to sodalite. In the glass-bonded sodalite waste form, the fission products are distributed between the glass and sodalite phases.

If an ion exchange process with zeolite-A has been used to selectively remove fission products from the salt, the resulting fission product loaded zeolite-A can be similarly converted into a glass-bonded sodalite ceramic waste form. Zeolite-A used for ion exchange is typically in pelletized form and must be milled to a fine particulate prior to blending with additional dried zeolite-4A and borosilicate glass. The flowsheet shown in Figure 17. 1 includes zeolite ion exchange followed by conversion of this zeolite into the ceramic waste form.

## 17. 3 Fluoride Volatility

In the fluoride volatility process, fluorine is reacted with the fuel. Fluorine is so much more reactive than even oxygen that small particles of ground oxide fuel will burst into flame when dropped into a chamber full of fluorine. This is known as flame fluorination; the heat produced helps the reaction proceed. Most of the uranium, which makes up the bulk of the fuel, is converted to uranium hexafluoride, the form of uranium used in uranium enrichment, which has a very low boiling point. Technetium, the main long-lived fission product, is also efficiently converted to its volatile hexafluoride. A few other elements also form similarly volatile hexafluorides, pentafluorides, or heptafluorides. The volatile fluorides can be separated from excess fluorine by condensation, then separated from each other by fractional distillation or selective reduction. Uranium hexafluoride and technetium hexafluoride have very similar boiling points and vapor pressures, which makes complete separation more difficult. Many of the fission products volatilized are the same ones volatilized in non-fluorinated, higher-temperature volatilization, such as iodine, tellurium and molybdenum; notable differences are that technetium is volatilized, but caesium is not.

Some transuranium elements such as plutonium, neptunium and americium can form volatile fluorides, but these compounds are not stable when the fluorine partial pressure is decreased. Most of the plutonium and some of the uranium will initially remain in ash which drops to the bottom of the flame fluorinator. The plutonium-uranium ratio in the ash may even approximate the composition needed for fast neutron reactor fuel. Further fluorination of the ash can remove all the uranium, neptunium, and plutonium as volatile fluorides; however, some other minor actinides may not form volatile fluorides and instead remain with the alkaline fission products. Some noble metals may not form fluorides at all, but remain in metallic form; however ruthenium hexafluoride is relatively stable and volatile. Distillation of the residue at higher temperatures can separate lower-boiling transition metal fluorides and alkali metal (Cs, Rb) fluorides from higher-boiling lanthanide and alkaline earth metal (Sr, Ba) and yttrium fluorides. The temperatures involved are much higher, but can be lowered somewhat by distilling in a vacuum. If a carrier salt like lithium fluoride or sodium fluoride is being used as a solvent, high-temperature distillation is a way to separate the carrier salt for reuse. Molten salt reactor designs carry out fluoride volatility reprocessing continuously or at frequent intervals. The goal is to return actinides to the molten fuel mixture for eventual fission, while removing fission products that are neutron poisons, or that can be more securely stored outside the reactor core while awaiting eventual transfer to permanent storage.

### 词汇

| | |
|---|---|
| vacuum furnace | 真空炉 |
| decontamination factor | 去污系数 |
| anode | 阳极 |
| electrolyte | 电解质 |
| cathode | 阴极 |
| cadmium | 镉 |
| electrode | 电极 |
| electrorefiner | 电解精炼 |
| oxide fuel | 氧化物燃料 |
| criticality limit | 临界极限 |

| | |
|---|---|
| zeolite | 沸石 |
| inert | 惰性的 |
| matrix material | 基体材料 |
| purification | 净化 |
| ion exchange | 离子交换 |
| adsorption | 吸附 |
| corrosive | (有)腐蚀性的 |
| fluoride volatility | 氟化挥发 |
| approximately | 接近于 |
| fast neutron reactor | 快中子反应堆 |
| distilling | 蒸馏 |
| alkaline earth metal | 碱土金属 |
| fluoride volatility reprocessing | 氟化物挥发性后处理 |
| neutron poison | 中子毒物 |
| reactor core | 堆芯 |
| storage | 储存;储存库 |

**注释**

1. Ceramic and metal waste streams are generated that immobilize fission products and, optionally, plutonium and minor actinides into high level waste forms.

产生的陶瓷和金属废物流将裂变产物以及一些钚和次锕系元素固定为高放废物形式。

2. The electrorefiner is at the centre of the flowsheet and is used to perform the primary separation of actinides from fission products.

电解精炼位于流程图的中心,它被用于从裂变产物中初步分离锕系元素。

3. Under normal conditions, Pu and minor actinides cannot deposit at the cathode, because their back-reaction with $UCl_3$ is thermodynamically spontaneous.

在正常条件下,钚和次锕系元素不能沉积在阴极上,因为它们与 $UCl_3$ 的逆反应在热力学上是自发的。

4. As current is passed between the anode and cathode, U metal is oxidized to $U^{3+}$ at the anode and reduced back to metallic form at the cathode.

当电流在阳极和阴极之间通过时,U 金属在阳极上被氧化为 $U^{3+}$,在阴极上又被还原为金属形式。

5. Fission product elements segregate between the anode basket and the molten salt during the electrorefining process. Noble metals such as Tc, Ru, and Rh remain with the cladding hulls in the anode basket.

在电解精炼过程中,裂变产物元素在阳极篮和熔盐之间分离。贵金属如 Tc、Ru 和 Rh 则保留在阳极篮中的包壳中。

6. Oxide fuel must first be converted to metallic form. This can be accomplished in an oxide

reduction step.

氧化物燃料首先必须被转化为金属形式。这可以通过氧化物还原步骤来完成。

7. This drying is used to maximize salt adsorption in the zeolite while minimizing evolution of water vapor in a high temperature, corrosive environment.

干燥被用于使沸石中的盐吸附达到最大化,同时使得高温、腐蚀性环境中水蒸气的析出达到最小化。

8. The goal is to return actinides to the molten fuel mixture for eventual fission, while removing fission products that are neutron poisons, or that can be more securely stored outside the reactor core while awaiting eventual transfer to permanent storage.

其目标是将锕系元素返回熔融燃料混合物中进行最终裂变,同时去除属于中子毒物的裂变产物,或在等待最终转移至永久性贮存库前可更安全地贮存在反应堆堆芯外的裂变产物。

# Chapter 18 Radioactive Waste Treatment

## 18.1 Source of Radioactive Waste

Radioactive waste comes from a number of sources. In countries with nuclear power plants, nuclear armament, or nuclear fuel treatment plants, the majority of waste originates from the nuclear fuel cycle and nuclear weapons reprocessing. Other sources include medical and industrial wastes, as well as naturally occurring radioactive materials (NORM) that can be concentrated as a result of the processing or consumption of coal, oil, gas, and some minerals, as discussed below.

**Nuclear fuel cycle**

Waste from the front-end of the nuclear fuel cycle is usually alpha-emitting waste from the extraction of uranium. It often contains radium and its decay products. Uranium dioxide ($UO_2$) concentrate from mining is a thousand or so times as radioactive as the granite used in buildings. It is refined from yellowcake ($U_3O_8$), then converted to uranium hexafluoride gas ($UF_6$). As a gas, it undergoes enrichment to increase the $^{235}U$ content from 0.7% to about 4.4% (LEU). It is then turned into a hard ceramic oxide ($UO_2$) for assembly as reactor fuel elements. The main by-product of enrichment is depleted uranium (DU), principally the $^{238}U$ isotope, with a $^{235}U$ content of about 0.3%. It is stored, either as $UF_6$ or as $U_3O_8$. Some is used in applications where its extremely high density makes it valuable such as anti-tank shells, and on at least one occasion even a sailboat keel. It is also used with plutonium for making mixed oxide fuel (MOX) and to dilute, or downblend, highly enriched uranium from weapons stockpiles which is now being redirected to become reactor fuel.

The back-end of the nuclear fuel cycle, mostly spent fuel rods, contains fission products that emit beta and gamma radiation, and actinides that emit alpha particles, such as uranium-234 (half-life 245 thousand years), neptunium-237 (2.144 million years), plutonium-238 (87.7 years) and americium-241 (432 years), and even sometimes some neutron emitters such as californium (half-life of 898 years for californium-251). These isotopes are formed in nuclear reactors.

**Nuclear weapons decommissioning**

Waste from nuclear weapons decommissioning is unlikely to contain much beta or gamma activity other than tritium and americium. It is more likely to contain alpha-emitting actinides such as Pu-239 which is a fissile material used in bombs, plus some material with much higher specific activities, such as Pu-238 or Po. In the past the neutron trigger for an atomic bomb tended to be beryllium and a high activity alpha emitter such as polonium; an alternative to polonium is Pu-238. For reasons of national security, details of the design of modern bombs are normally not released to the open literature. Some designs might contain a radioisotope thermoelectric generator using Pu-238 to provide a long-lasting source of electrical power for the electronics in the device. It is likely that the fissile material of an old bomb which is due for refitting will contain decay products of the plutonium isotopes used in it, these are likely to include U-236 from Pu-240 impurities, plus some U-235 from decay of the Pu-239; due to the relatively long half-life of these Pu isotopes, these wastes from radioactive decay of bomb core material would be very small, and in any case, far less dangerous (even in terms of simple radioactivity) than the Pu-239 itself. The beta decay of Pu-241 forms Am-241; the in-growth of americium is likely to be a greater problem than the decay of Pu-239 and Pu-240 as the americium is a gamma emitter (increasing external-exposure to workers) and is an alpha emitter which can cause the generation of heat. The plutonium could be separated from the americium by several different processes; these would include pyrochemical processes and aqueous/organic solvent extraction. A truncated PUREX type extraction process would be one possible method of making the separation. Naturally occurring uranium is not fissile because it contains 99.3% of U-238 and only 0.7% of U-235.

**Medicine**

Radioactive medical waste tends to contain beta particle and gamma ray emitters. It can be divided into two main classes. In diagnostic nuclear medicine, a number of short-lived gamma emitters such as technetium-99m are used. Many of these can be disposed of by leaving them to decay for a short time before disposal as normal waste. Other isotopes used in medicine, with half-lives in parentheses, include:

- Y-90, used for treating lymphoma (2.7 d)
- I-131, used for thyroid function tests and for treating thyroid cancer (8 d)
- Sr-89, used for treating bone cancer, intravenous injection (52 d)
- Ir-192, used for brachytherapy (74 d)
- Co-60, used for brachytherapy and external radiotherapy (5.3 a)
- Cs-137, used for brachytherapy and external radiotherapy (30 a)
- Tc-99, product of the decay of Technetium-99m (221 000 a)

## 18.2 Classification of Radioactive Waste

Classification of radioactive waste varies by country. The IAEA, which publishes the *Radioactive Waste Safety Standards (RADWASS)*, also plays a significant role. The proportion of various types of waste generated in the UK:

- 94%——low-level waste (LLW)
- 6%——intermediate-level waste (ILW)
- <1%——high-level waste (HLW)

LLW is generated from hospitals and industry, as well as the nuclear fuel cycle. Low-level wastes include paper, rags, tools, clothing, filters, and other materials which contain small amounts of mostly short-lived radioactivity. Materials that originate from any region of an active area are commonly designated as LLW as a precautionary measure even if there is only a remote possibility of being contaminated with radioactive materials. Such LLW typically exhibits no higher radioactivity than one would expect from the same material disposed of in a non-active area, such as a normal office block. Example LLW includes wiping rags, mops, medical tubes, laboratory animal carcasses, and more. LLW waste makes 94% of all radioactive waste volume in the UK.

Some high-activity LLW requires shielding during handling and transport but most LLW is suitable for shallow land burial. To reduce its volume, it is often compacted or incinerated before disposal.

ILW contains higher amounts of radioactivity compared to low-level waste. It generally requires shielding, but not cooling. Intermediate-level wastes includes resins, chemical sludge and metal nuclear fuel cladding, as well as contaminated materials from reactor decommissioning. It may be solidified in concrete or bitumen or mixed with silica sand and vitrified for disposal. As a general rule, short-lived waste (mainly non-fuel materials from reactors) is buried in shallow repositories, while long-lived waste (from fuel and fuel reprocessing) is deposited in geological repository. Regulations in the United States do not define this category of waste; the term is used in Europe and elsewhere. ILW makes 6% of all radioactive waste volume in the UK.

HLW is produced by nuclear reactors and the reprocessing of nuclear fuel. The exact definition of HLW differs internationally. After a nuclear fuel rod serves one fuel cycle and is removed from the core, it is considered HLW. Spent fuel rods contain mostly uranium with fission products and transuranic elements generated in the reactor core. Spent fuel is highly radioactive and often hot. HLW accounts for over 95% of the total radioactivity produced in the process of nuclear electricity generation but it contributes to less than 1% of volume of all radioactive waste produced in the UK.

## 18.3 Treatment of Aqueous Waste

The aim of the radioactive waste treatment is to minimize the volume of waste requiring management. Treatment process selection for waste depends upon its radiological and physicochemical properties and the quantity.

The processes available for treating aqueous radioactive waste are mainly: ion exchange/sorption, chemical precipitation, evaporation or ultrafiltration/reverse osmosis. However, liquid containing suspended matter must be treated to remove the particulates before primary treatment or after it. Sedimentation, decantation, filtration or centrifugation are treatments used commonly to clear the effluent wastes or to remove miscellaneous debris or insoluble particles.

Chemical precipitation processes are regularly used for removing radioactivity from low and intermediate level aqueous wastes at fuel reprocessing facilities, research laboratories and power stations. Precipitation processes are greatly versatile, relatively low investment and operational costs; and may treat from large volumes of liquid effluents containing relatively low concentrations of active species to those containing large amounts of particulates or high concentrations of inactive salts. However, in some cases, a pretreatment stage, such as oxidization of organic contaminants, decomposition of complexed species, pH adjustment, change of the valency state or adjustment of the ionic species, should be applied prior to the formation of precipitate in order to improve the process. Radionuclides can be removed by precipitation, co-precipitation with a carrier or sorption on to particulates present in the waste.

Ion exchange methods have extensive applications to remove soluble radionuclides from liquid waste produced in nuclear fuel cycle operations, radioisotope production and research facilities. It is very effective at transferring the radioactive content of a large volume of liquid into a small volume of solid. Ion exchange process involves the replacement of cations or anions between an insoluble solid matrix containing ionizable polar groups and a liquid solution. When the ionic groups are negatives, the exchange will involve cations and when they are positively charged they involve anions. The process is selective, stoichiometric and, as a rule, reversible; therefore ion exchangers can be "regenerated" and radioactive liquid waste recovered with high activity content or if the exchangers become "exhausted" they are removed and treated as radioactive wastes. A wide range of materials is available for the ion exchange treatment of radioactive liquids: (a) natural ion exchangers (clays, zeolites, cellulose, charcoals, collagen) and (b) synthetic materials such as zeolites, hydrous oxide gels of metals or organic resins formed by highly polymerized cross-linked hydrocarbons containing ionic groups (sulfonic acid, carboxylic acid, amino groups, etc.). Ion exchange processes can be operated in batch or continuous modes and if the wastes contain high concentrations of salts, suspended solids, organic contaminants or the radionuclide ionic form not suitable, the liquid wastes will have to be pretreated before exchange process.

Evaporation process is effective for concentrating or removing salts, heavy metals and a variety of hazardous materials from waste effluent, reducing large volumes of liquid wastes with high factor decontamination. The process is commonly used for the treatment of high, intermediate and low level waste effluents; in particular for the treatment of small volumes of highly active effluents and may be carried out through the use of commercially available evaporation equipment. However, evaporation has some important limitations: unsuitable for waste effluents containing large concentrations of inactive salts, expensive because its large energy requirement and the presence of some organic compounds can produce explosions during evaporation.

## 18.4 Treatment of Radioactive Organic Liquid

Liquid scintillation, solvents, oils and diverse biological fluids, generated in nuclear research centres, medical centres or industries are considered as radioactive organic liquid wastes. These wastes may present radioactive and chemical or biochemical hazards requiring treatments to remove or destroy chemically or biochemically hazardous components. The objective is to reduce the volume of radioactive waste which requires storage, transport, conditioning and land disposal, eliminating the organic components to enhance compatibility of the treated waste with secondary conditioning processes. Processes such as incineration, wet oxidation, acid digestion, electroche-mical oxidation and distillation, can be applied for treating radioactive organic liquid.

Incineration is used for reduction of solid and liquid radioactive waste volume, downscaling land requirements for disposal. Incineration combusts or oxidizes wastes at high temperatures, generating as end products of the complete incineration: $CO_2$, $H_2O$, $SO_2$, NO, and HCl gases. Emission control equipment for particulates, $SO_2$, $NO_x$ and products of incomplete oxidation are needed to control emissions of regulated air pollutants. The disadvantages of radioactive waste treatment with incineration are: off-gas filtering systems are required to control radioactive discharges, thickening and dewatering wastes pretreatment may be required, is not economical for small solid waste plants and secondary waste volumes may be large.

The organic components of radioactive wastes such as ion exchange resins, foams, cellulosic waste and liquid scintillation can be transformed, degraded, or immobilized using wet oxidation. Wet oxidation remedial process involves injecting an oxidizing agent, such as hydrogen peroxide, activated sodium persulfate, ozone, Fenton's Reagent (hydrogen peroxide with an iron catalyst) or other oxidant into the subsurface to destroy organic compounds. The complete mineralization of carbon-based compound wastes by most chemical oxidizers produces carbon dioxide, water, and oxygen as well as minor concentrations of nontoxic ions, salts, and acids. Wet oxidation is thus a process analogous to incineration, with the advantage of using low temperatures.

Acid digestion is an oxidative destruction technology for some liquid organic wastes (hexane, TBP) and organic constituents of mixed waste such as cellulose (paper), polyethylene, latex rubber, Tyvek™, neoprene, polyvinylchloride, polystyrene ion exchange resins, filters, plastics, and/or chlorinated cutting oils organic, that may reduce the waste volume of 20 to 100 times. Acid digestion process uses a mixed of nitric acid in a phosphoric acid carrier solution at temperatures below 200 ℃ and at atmospheric or moderate pressures (<20 psig). The principal organic portion of the waste is broken down and mineralized by the acid solution producing: inorganic constituents in solution, which can be immobilized easily in a glass or ceramic, and gases ($CO_2$, CO, $O_2$ and $NO_x$) that can be treated in an off-gas scrubbing system, to convert $NO_x$ to reusable nitric acid.

Distillation is a radioactive waste volume reduction technique used for pretreating liquid scintillation and miscellaneous solvent waste in conventional equipment. The process is simple, known, and cost effective if the valuable solvent is recycled or reused. The active residue could be either immobilized or destroyed by incineration.

## 18.5 Treatment of Solid Waste

Solid wastes are produced by all applications and uses of radioactive materials, in normal operations and maintenance activities. Solid, low and intermediate level wastes are generally segregated into combustible, compactible and non-compactible forms. Treatments for solid waste are used to reduce the waste volume and/or convert the waste into a form suitable for handling, storage and disposal.

Decontamination is defined as the removal of contamination from areas or surfaces of facilities or equipment by washing, heating, chemical or electrochemical action, mechanical cleaning or by other means. The decontamination objectives are mainly: to reduce the volume of equipment and materials requiring storage and disposal in licensed disposal facilities, to remove contamination from components or systems, to reduce dose levels in the installations and to restore sites and facilities to an unconditional-use condition. Decontamination processes may divide into chemical, electrochemical and mechanical processes.

- Chemical decontamination. The chemical decontamination uses concentrated or dilute chemical reagents in contact with the contaminated item to dissolve the contamination layer, covering the base metal and eventually a part of the base metal.
- Decontamination by melting presents the particular advantage of homogenising a number of radionuclides in the ingots and concentrating other radionuclides in the slag and filter dust resulting from the melting process, thus decontaminating the primary material. The problem with inaccessible surfaces or complex geometries is eliminated and the remaining radioactivity content is homogenised over the total mass of the ingot.
- Mechanical and manual decontamination include wet or dry abrasive blasting, grinding of surfaces and removal of concrete by spalling or scarifying, washing, swabbing, foaming agents, and latex-peelable coatings. These techniques are most applicable to the decontamination of structural surfaces which may be cleaned by sweeping, wiping, scrubbing or removed by grit blasting, scarifying, drilling and spalling.

A wet abrasive-blasting system uses a combination of water, abrasive media and compressed air, and is normally applied in 24 a self-contained, leak tight, stainless steel enclosure. The dry abrasive-blasting technique, commonly called sandblasting or abrasive jetting, uses abrasive materials suspended in a medium that is projected onto the surface being treated, resulting in a uniform removal of surface contamination. The scarification process removes the top layers of a contaminated surface down to the depth of the sound, uncontaminated surface. There are two basic disadvantages with the mechanical methods: the surface of the workpiece has to be accessible and many methods may produce airborne dust.

Compaction: Compaction is performed in order to reduce the waste volume and concentrate the radionuclides. Plastics, paper, absorbent material, and cloth are compatible in conventional compactors. Metal pipe, valves, conduit, wood, and other like items are compatible in super compactors. Compactors can range from low-force compaction systems (5 t or more) through to presses with a compaction force over 1 000 t (super compactors). Volume reduction factors are typically between 3 and 10, depending on the waste material being treated.

Cutting: Cutting and sawing operations are carried out mainly on large items which consist usually of metals or plastics. The waste has to be reduced in size to make it fit into packaging containers or to submit it to treatment such as incineration. The cutting is carried out either in the dry state in cells, using remote control when necessary and with conventional tools, or underwater. The cutting may also be done with plasma-jets, laser torches, or explosive fuses.

Crushing: Crushing techniques may be used for size reduction of friable solids (e.g., glass, concrete, ceramics). In principle, all types of mill, grinder, and crushing machines of conventional technology can be used.

Shredding: Shredding reduces void space and is particularly effective when plastics are compacted. Air, which is trapped between the folds of bulk plastic and in plastic bags and sleeving, takes up storage space. When the plastic is shredded, better use is made of the waste container space.

Incineration: The size reduction, mixing and blending of the solid wastes are necessary for successful combustion operation.

### 词汇

| | |
|---|---|
| naturally occurring radioactive materials (NORM) | 天然放射性物质 |
| front-end of the nuclear fuel cycle | 核燃料循环前端 |
| back-end of the nuclear fuel cycle | 核燃料循环后端 |
| depleted uranium | 贫化铀 |
| brachytherapy | 近距离放射治疗 |
| reactor decommission | 反应堆退役 |
| bitumen | 沥青 |
| silica | 二氧化硅 |
| vitrify | 玻璃化 |
| repository | 仓库 |
| ion exchange/sorption | 离子交换/吸附 |
| chemical precipitation | 化学沉淀 |
| ultrafiltration/reverse osmosis | 超滤/反渗透 |
| decantation | 倾析 |
| miscellaneous debris | 杂项碎屑 |
| zeolites | 沸石 |
| cross-linkcd | 交联 |
| liquid scintillation | 液体闪烁 |
| incineration | 焚化 |
| wet oxidation | 湿法氧化 |
| acid digestion | 酸消化法 |
| distillation | 蒸馏 |
| homogenise | 使均质 |
| ingot | 铸块 |
| slag | 炉渣；矿渣 |

| | |
|---|---|
| abrasive blasting | 喷砂 |
| grind | 磨 |
| scarify | 翻松 |
| swab | 擦洗 |
| foaming agent | 发泡剂 |
| compaction | 压实 |
| cutting | 切割 |
| friable solids | 易碎固体 |
| shredding | 粉碎 |

## 注释

1. The back-end of the nuclear fuel cycle, mostly spent fuel rods, contains fission products that emit beta and gamma radiation, and actinides that emit alpha particles, such as uranium-234 (half-life 245 thousand years), neptunium-237(2.144 million years), plutonium-238(87.7 years) and americium-241(432 years), and even sometimes some neutron emitters such as californium (half-life of 898 years for californium-251). These isotopes are formed in nuclear reactors.

核燃料循环的后端,主要是乏燃料棒,包含发射 β 粒子和 γ 粒子的裂变产物,以及发射 α 粒子的锕系元素,如 $^{234}U$(半衰期 24.5 万年)、$^{237}Np$(214.4 万年),$^{238}Pu$(87.7 年)和 $^{241}Am$(432 年),甚至有时还有一些中子发射体,如锎($^{251}Cf$ 的半衰期为 898 年)。这些同位素是在核反应堆中形成的。

2. Some high-activity LLW requires shielding during handling and transport but most LLW is suitable for shallow land burial. To reduce its volume, it is often compacted or incinerated before disposal.

一些高活性低放废物在搬运和运输过程中需要屏蔽,但大多数低放废物适用于浅埋。为了减少其体积,通常在处理前对其进行压实或焚烧。

3. However, in some cases, a pretreatment stage, such as oxidization of organic contaminants, decomposition of complexed species, pH adjustment, change of the valency state or adjust the ionic species, should be applied prior to the formation of precipitate in order to improve the process.

然而,在某些情况下,为了改进工艺,应在形成沉淀物之前进行预处理,如有机污染物的氧化、络合物种的分解、pH 调节、价态的改变或离子物种的调节。

4. The objective is to reduce the volume of radioactive waste which requires storage, transport, conditioning and land disposal, eliminating the organic components to enhance compatibility of the treated waste with secondary conditioning processes.

目标是减少需要储存、运输、处理和土地处置的放射性废物的体积,消除有机成分,以增强处理后的废物与二次处理过程的兼容性。

5. The process is selective, stoichiometric and, as a rule, reversible; therefore ion exchangers can be "regenerated" and radioactive liquid waste recovered with high activity content or if the exchangers become "exhausted" they are removed and treated as radioactive wastes.

该过程是选择性的、化学计量的,通常是可逆的;因此,离子交换剂可以“再生”,并回收具有高活性含量的放射性液体废物,或者如果交换剂“耗尽”,则将其作为放射性废物进行移除和处理。

6. The decontamination objectives are mainly: to reduce the volume of equipment and materials requiring storage and disposal in licensed disposal facilities, to remove contamination from components or systems, to reduce dose levels in the installations and to restore sites and facilities to an unconditional-use condition.

去污目标主要是:减少需要在有许可证的处置设施中储存和处置的设备和材料的体积,清除部件或系统的污染,降低装置中的剂量水平,并使现场和设施恢复到无条件使用状态。

7. Decontamination by melting presents the particular advantage of homogenising a number of radionuclides in the ingots and concentrating other radionuclides in the slag and filter dust resulting from the melting process, thus decontaminating the primary material.

熔化去污具有使铸锭中的许多放射性核素均匀化,并将熔化过程中产生的炉渣和过滤尘中的其他放射性核素浓缩的特殊优势,从而对原材料进行去污。

# Chapter 19 Radioactive Waste Disposal

## 19.1 General Description

Most low-level radioactive waste (LLRW) is typically sent to land-based disposal immediately following its packaging for long-term management. This means that for the majority (about 90% by volume) of all of the waste types produced by nuclear technologies, a satisfactory disposal means has been developed and is being implemented around the world.

For used fuel designated as high-level radioactive waste (HLRW), the first step is storage to allow decay of radioactivity and heat, making handling much safer. Storage of used fuel may be in ponds or dry casks, either at reactor sites or centrally. Beyond storage, many options have been investigated which seek to provide publicly acceptable, safe, and environmentally sound solutions to the final management of radioactive waste. The most widely favored solution is deep geological disposal. The focus is on how and where to construct such facilities.

Used fuel that is not intended for direct disposal may instead be reprocessed in order to recycle the uranium and plutonium it contains. Some separated liquid HLRW arises during reprocessing; this is vitrified in glass and stored pending final disposal.

Intermediate-level radioactive waste (ILRW) that contains long-lived radioisotopes is also stored pending disposal in a geological repository. In the USA, defence-related transuranic (TRU) waste — which has similar levels of radioactivity to some ILRW — is disposed of in the Waste Isolation Pilot Plant (WIPP) deep geological repository in New Mexico. A number of countries dispose of ILRW containing short-lived radioisotopes in near-surface disposal facilities, as used for LLRW disposal.

Some countries are at the preliminary stages of their consideration of disposal for ILRW and HLRW, whilst others, such as Finland and Sweden, have made good progress. Finland's Onkalo repository is expected to start operating soon. It will be the first deep geological repository licensed for the disposal of used fuel from civil reactors. When considering these, it should be noted that the suitability of an option or idea is dependent on the waste form, volume, and radioactivity of the waste. As such, waste management options and ideas described in this section are not all applicable to different types of waste.

## 19.2 Near-surface Disposal and Deep Geological Disposal

The International Atomic Energy Agency (IAEA) definition of this option is the disposal of waste, with or without engineered barriers, in:

Near-surface disposal facilities at ground level. These facilities are on or below the surface where the protective covering is of the order of a few meters thick. Waste containers are placed in constructed vaults and when full the vaults are backfilled. Eventually, they will be covered and capped with an impermeable membrane and topsoil. These facilities may incorporate some form of drainage and possibly a gas venting system.

Near-surface disposal facilities in caverns below ground level. Unlike near-surface disposal at ground level, where the excavations are conducted from the surface, shallow disposal requires underground excavation of caverns. The facility is at a depth of several tens of metres below the Earth's surface and accessed through a drift.

The term near-surface disposal replaces the terms "shallow land" and "ground disposal", but these older terms are still sometimes used when referring to this option. These facilities will be affected by long-term climate changes (such as glaciation) and this effect must be taken into account when considering safety, as such changes could disrupt these facilities. This type of facility is therefore typically used for LLW and short-lived ILW with half-lives of up to 30 a.

The long timescales over which some waste remains radioactive has led to the idea of deep disposal in underground repositories in stable geological formations. Isolation is provided by a combination of engineered and natural barriers (rock, salt, clay) and no obligation to actively maintain the facility is passed on to future generations. This is often termed a "multi-barrier" concept, with the waste packaging, the engineered repository, and the geology all providing barriers to prevent the radionuclides from reaching humans and the environment. In addition, deep groundwater is generally devoid of oxygen, minimizing the possibility of chemical mobilization of waste.

Deep geological disposal is the preferred option for nuclear waste management in most countries, including Argentina, Australia, Belgium, Canada, China, Czech Republic, Finland, France, Japan, the Netherlands, Republic of Korea, Russia, Spain, Sweden, Switzerland, the UK, and the USA. Hence, there is much information available on different disposal concepts; a few examples are given here. The only purpose-built deep geological repository that is currently licensed for disposal of nuclear material is the WIPP in the USA, but it does not have a license for disposal of used fuel or HLW. Plans for disposal of spent fuel are particularly well advanced in Finland, as well as Sweden, France, and the USA, though in the USA there have been political delays. In Canada and the UK, deep disposal has been selected and the site selection processes have commenced.

## 19.3 Mined Repositories

The most widely proposed deep geological disposal concept is for a mined repository comprising tunnels or caverns into which packaged waste would be placed. In some cases (e. g. wet rock) the waste containers are then surrounded by a material such as cement or clay (usually bentonite) to provide another barrier (called buffer and/or backfill). The choice of waste container materials and design, as well as the buffer/backfill material varies depending on the type of waste to be contained and the nature of the host rock-type available.

Excavation of a deep underground repository using standard mining or civil engineering technology is limited to accessible locations (e. g., under land or nearshore), to rock units that are reasonably stable and without major groundwater flow, and to depths of between 250 m and 1 000 m. The contents of the repository would be retrievable in the short term, and if desired, longer-term.

The Swedish proposed KBS-3 disposal concept uses a copper container with a steel insert to contain the spent fuel. After placement in the repository about 500 meters deep in the bedrock, the container would be surrounded by a bentonite clay buffer to provide a very high level of containment of the radioactivity in the spent fuel over a very long time period (shown in Figure 19. 1). In June 2009, the Swedish Nuclear Fuel and Waste Management Company (SKB) announced its decision to locate the repository at Östhammar (Forsmark). Finland's repository programme is also based on the KBS-3 concept. Spent nuclear fuel packed in copper canisters will be embedded in the Olkiluoto bedrock at a depth of around 400 meters. The country's nuclear waste management company, Posiva Oy, expects the repository to begin disposal operations soon. Its construction was licensed in November 2015. The deposits of native (pure) copper in the world have proved that the copper used in the final disposal container can remain unchanged inside the bedrock for extremely long periods, if the geochemical conditions are appropriate (low levels of groundwater flow). The findings of ancient copper tools, many thousands of years old, also demonstrate the long-term corrosion resistance of copper, making it a credible container material for long-term radioactive waste storage.

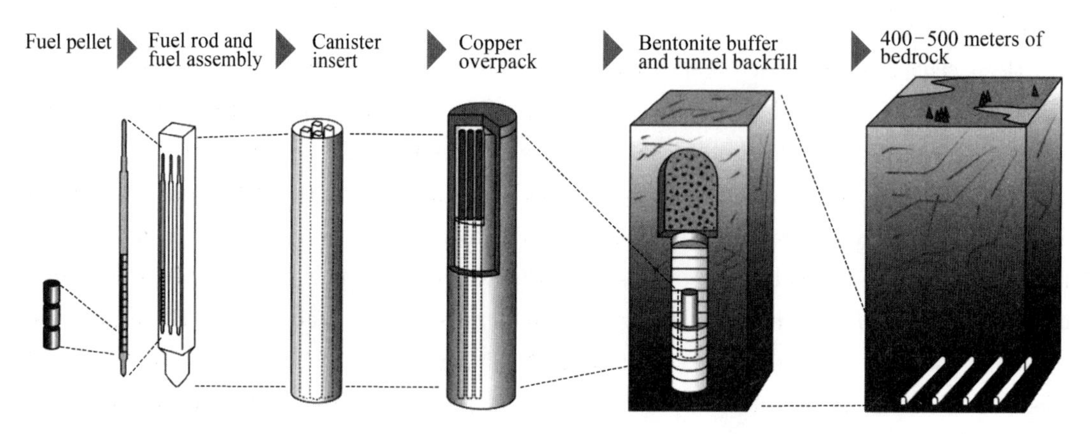


**Figure 19. 1 Multi-barrier disposal concept**

## 19. 4 The Multi-barrier Approach

Geological disposal provides safety through a combination of man-made (engineered) and natural barriers that work together to provide isolation and containment of radioactive waste. This is called the multi-barrier approach. Isolation means putting the waste well away from people and the surface environment. It is provided by:

The depth of a GDF, which will be between 200 and 1 000 meters below the surface in stable rock. Deep underground, the waste will be much better protected in the event of earthquakes, tsunamis and long-term environmental changes such as future ice ages.

The choice of a site away from known areas of significant underground resources such as fossil fuels or minerals are scarce. This reduces the likelihood of humans disturbing the facility in the future.

Containment means keeping the waste where we put it. It is provided by:

The solid form of the radioactive waste itself.

The long-lasting metal or concrete waste container.

The material placed immediately around the waste containers to add further protection. We call this buffer or backfill.

Other engineered features of a GDF such as the seals in tunnels or vaults.

The stable underground environment in which the facility is built.

Unlike other hazardous wastes, radioactive waste gets less dangerous with time. This is because it undergoes radioactive "decay", i. e., it gives off radiation and eventually becomes a stable (non-radioactive) material. It will take many tens or even hundreds of thousands of years for radioactive waste to decay to harmless levels. We therefore engineer barriers that will work together with the natural barrier of the rock to provide safety. We also think about how the environment around a GDF might change over such a long timescale. The barriers in a multi-barrier approach include:

(1) The waste form

Only solid waste will be sent to a GDF. HLRW, a liquid by-product from reprocessing nuclear reactor fuel, is transformed into blocks of glass for storage and disposal. Other wastes are mixed with cement or sometimes are set in a resin matrix.

(2) The waste container

The waste is put into containers made of concrete or a metal such as cast iron, stainless steel or copper. After a GDF is closed, containers provide a physical barrier that prevents or limits the release of radioactivity. A container may be designed to last for anything from a few hundred to many tens of thousands of years, depending on the materials used and the thickness of the container walls. Some radioactive waste has already been placed in containers, ready for disposal, while other waste has yet to be packaged.

(3) A buffer or backfill

The space between the waste containers and the rock walls of a GDF will be filled with bentonite (a type of clay), cement or crushed rock. This protects the containers and prolongs their life. No container will last forever, but when containers do eventually fail, this barrier will slow down the release of radioactivity out of the GDF.

(4) Seals

Before a GDF is closed, all the excavated space will be backfilled and access ways will be sealed to prevent human access. Far into the future, when other engineered barriers have degraded, the seals will limit the movement of radioactivity along the tunnels and shafts.

(5) The rock barrier

Placing radioactive waste deep underground puts it far beyond people's reach, so that it is safe and secure. The rock will shield people from the radiation and, depending on the rock type, will either limit or completely prevent radioactivity from moving towards the surface when other barriers eventually degrade. Disposal deep underground will also ensure that the waste can never be exposed at the surface even in the event of a change in sea level or future ice ages.

## 19.5 Deep Boreholes

As well as mined repositories, which have been the focus of most international efforts so far, deep borehole disposal has been considered as an option for geological isolation for many years, including original evaluations by the US National Academy of Sciences in 1957 and more recent conceptual evaluations. In contrast to recent thinking on mined repositories, the contents would not be retrievable. The concept consists of drilling a borehole into basement rock to a depth of up to about 5 000 meters, emplacing waste canisters containing used nuclear fuel or vitrified radioactive waste from reprocessing in the lower 2 000 meters of the borehole, and sealing the upper 3 000 meters of the borehole with materials such as bentonite, asphalt or concrete. The disposal zone of a single borehole could thus contain 400 steel canisters each 5 meters long and one-third to half a meter in diameter. The waste containers would be separated from each other by a layer of bentonite or cement. Boreholes can be readily drilled offshore as well as onshore in both crystalline and sedimentary host rocks. This capability significantly expands the range of locations that can be considered for the disposal of radioactive waste. Deep borehole concepts have been developed (but not implemented) in several countries, including Denmark, Sweden, Switzerland, and the USA. Compared with deep geological disposal in a mined underground repository, placement in deep boreholes is considered to be more expensive for large volumes of waste. This option was abandoned in countries such as Sweden, Finland, and the USA, largely on economic grounds. The borehole concept remains an attractive proposition for the disposal of smaller waste forms including sealed radioactive sources from medical and industrial applications. An October 2014 U. S. Department of Energy (DOE) report said: "Preliminary evaluations of deep borehole disposal indicate a high potential for robust isolation of the waste, and the concept could offer a pathway for earlier disposal of some wastes than might be possible in a mined repository." In January 2016, the DOE commissioned a team led by Battelle to drill a 4 880-metre test borehole into crystalline basement rock in North Dakota, but the project was later scrapped following local opposition.

**词汇**

disposal 处理

deep geological disposal 深层地质处理

direct disposal 直接处理

| | |
|---|---|
| ILRW | 中级放射性废物 |
| pending | 待定的 |
| Waste Isolation Pilot Plant (WIPP) | 废弃物隔离试验厂 |
| near-surface disposal facilities | 近地表处理设施 |
| civil reactors | 民用反应堆 |
| waste form | 废物形态 |
| vault | 储藏室 |
| membrane | 膜 |
| underground repositories | 地下处置库 |
| multi-barrier concept | 多重屏障概念 |
| devoid | 缺乏 |
| mobilization | 移动 |
| nuclear waste management | 核废料管理 |
| buffer/backfill | 缓冲区/回填区 |
| excavation | 挖掘 |
| depths | 深度 |
| corrosion | 腐蚀性 |
| combination | 结合 |
| mineral | 矿物 |
| scarce | 稀少的 |
| container | 容器 |
| hazardous | 危险的 |
| waste container | 废物箱 |
| stainless steel | 不锈钢 |
| seal | 密封 |
| rock barrier | 岩石屏障 |
| borehole | 钻孔 |
| canister | 罐 |
| diameter | 外径 |
| crystalline | 结晶状 |
| sedimentary | 沉积形式的 |

## 注释

1. In the USA, defence-related transuranic (TRU) waste — which has similar levels of radioactivity to some ILRW — is disposed of in the waste isolation pilot plant (WIPP) deep geological repository in New Mexico.

在美国,与国防有关的超铀(TRU)废物——其放射性水平与某些ILRW类似——被处置在新墨西哥州的废物隔离试验厂(WIPP)深地质处置库中。

2. Eventually they will be covered and capped with an impermeable membrane and topsoil. These facilities may incorporate some form of drainage and possibly a gas venting system.

最后,将用防渗膜和表土对其进行覆盖和封顶。这些设施可能包括某种形式的排水系统,也可能包括气体排放系统。

3. Unlike near-surface disposal at ground level, where the excavations are conducted from the surface, shallow disposal requires underground excavation of caverns.

与在地面进行挖掘的近地表处置不同,浅层处置需要在地下挖掘岩洞。

4. This is often termed a "multi-barrier" concept, with the waste packaging, the engineered repository, and the geology all providing barriers to prevent the radionuclides from reaching humans and the environment.

这通常被称为"多屏障"概念,废物包装、工程处置库和地质都是防止放射性核素进入人类和环境的屏障。

5. Excavation of a deep underground repository using standard mining or civil engineering technology is limited to accessible locations (e. g., under land or nearshore), to rock units that are reasonably stable and without major groundwater flow, and to depths of between 250 m and 1 000 m.

使用标准采矿或土木工程技术挖掘深层地下处置库仅限于可进入的地点(如陆地下或近岸)、相当稳定且无主要地下水流的岩石单元,以及250米至1 000米的深度。

6. After placement in the repository about 500 meters deep in the bedrock, the container would be surrounded by a bentonite clay buffer to provide a very high level of containment of the radioactivity in the spent fuel over a very long time period.

容器放置在约500米深的基岩中的处置库后,周围将有膨润土缓冲,以便在很长一段时间内对乏燃料中的放射性进行高度密封。

7. Geological disposal provides safety through a combination of man-made (engineered) and natural barriers that work together to provide isolation and containment of radioactive waste.

地质弃置通过人工(工程)屏障和天然屏障的结合,共同隔离和遏制放射性废物,从而确保安全。

8. Far into the future, when other engineered barriers have degraded, the seals will limit the movement of radioactivity along the tunnels and shafts.

在遥远的未来,当其他工程屏障发生退化时,密封装置将限制放射性物质沿隧道和竖井移动。

9. The rock will shield people from the radiation and, depending on the rock type, will either limit or completely prevent radioactivity from moving towards the surface when other barriers eventually degrade.

岩石将使人们免受辐射的影响,而且根据岩石类型的不同,当其他屏障最终降解时,岩石将限制或完全阻止放射性物质向地表移动。

10. As well as mined repositories, which have been the focus of most international efforts so far, deep borehole disposal has been considered as an option for geological isolation for many

years, including original evaluations by the US National Academy of Sciences in 1957 and more recent conceptual evaluations.

迄今为止,矿物处置库一直是国际努力关注的重点。与此同时,深钻孔处置多年来一直被认为是地质隔离的一种选择,包括 1957 年美国国家科学院的最初评估和最近的概念性评估。

# Chapter 20 Nuclear Safety

## 20.1 Radioactive Releases and Possible Effects

The main objection against nuclear power is the risk of release of radionuclides to the environment where it may cause health effects in humans. Such releases occur from mining and milling operations, particularly of uranium ores, from the nuclear fuel fabrication processes, from normal operation of nuclear reactors, from reprocessing of spent nuclear fuel, from nuclear weapons production and recovery, from transportation of nuclear material, from testing of nuclear weapons and accidents, and from storage of nuclear wastes. Table 20.1 compares the estimated collective dose to the public from the various activities of the nuclear fuel cycle, from mining and milling through nuclear waste disposal; the significant point is the recognition that the mining operation for uranium is the major contributor in the dose to the global public by quite a large fraction.

**Table 20.1 Collective effective dose to the public from radionuclides released in effluents from the nuclear fuel cycle 1995—1997 (UNSCEAR 2000)**

| Source | Normalized collective effective dose (manSv/GWey) |
|---|---|
| | Local and regional effects |
| Mining | 0.19 |
| Milling | 0.008 |
| Mine and mill tailings (releases over 5 a) | 0.04 |
| Fuel fabrication | 0.003 |
| Reactor operation | |
| Atmospheric | 0.4 |
| Aquatic | 0.04 |
| Reprocessing | |
| Atmospheric | 0.04 |
| Aquatic | 0.09 |
| Transportation | <0.1 |

**Table 20.1** (Continued)

| Source | Normalized collective effective dose (manSv/GWey) |
|---|---|
| | Solid wate disposal and global effects |
| Mine and mill tailings | |
| Releases of radon over 10 000 a | 7.5 |
| Reactor operation | |
| Low-level waste disposal | 0.000 05 |
| Intermediate-level waste disposal | 0.5 |
| Reprocessing solid waste disposal | 0.05 |
| Globally dispersed radionuclides (truncated to 10 000 a) | 40 |

## 20.2 Radionuclides of Environmental Concern

Most of the radionuclides produced in nuclear tests, in accidents and in the normal fuel cycle are short-lived. In Table 20.2, longer lived fission products and activation products from these systems are listed as these represent the major concern to the general public if they are allowed to enter the environment; we exclude nuclides with insignificant contribution to the total activity after 10 a. In addition, several heavy radionuclides are formed by neutron capture reactions such as $^{236}U$, $^{237}Np$, $^{238-242}Pu$, $^{241}Am$, etc.

To assess the potential for these radionuclides to cause harm to humans, their geochemical and biological behavior must be evaluated. For example, since Kr is a chemically inert gaseous element, it would have little effect on a person who inhaled and immediately exhaled the small amount which might be present in the air. By contrast, other nuclides with high activity, $^{90}Sr-^{90}Y$ and $^{137}Cs$-$^{137m}Ba$, have active geological and biological behavior and can present much more significant radiation concerns to humans. In normal operations, these nuclides would of course be released in quite insignificant amounts, as represented by the value for aquatic releases in reactor operation and reprocessing. Similarly, the heavy elements (Np, Pu, Am) would not be released in normal operation at levels that would be of concern. We have pointed out that some low level releases as a result of normal operations are allowed by the health authorities, who also monitor these levels. In the effluents from reprocessing plants, the relatively long-lived nuclides such as $^{3}H$, $^{14}C$, $^{85}Kr$, $^{99}Tc$ and $^{129}I$ are of major concern. The liquid effluents from nuclear power plants and from reprocessing plants are about equally responsible for the global collective dose commitment of nuclear power generation.

**Table 20.2 Some long-lived radionuclides produced in fission**

| Nuclide | Half-life (a)† | Decay; $\beta^-$ (energy MeV)‡ | Thermal fission yield(%)†† | Activity (TBq/t U) |
|---|---|---|---|---|
| $^{79}Se$ | $<6.5\times10^4$ | 0.150 9 | 0.044 3 | 251 |
| $^{85}Kr$ | 10.72 | 0.687 4 | 1.318 | 1 930 |
| $^{87}Rb$ | $4.8\times10^{10}$ | 0.282 3 | 2.558 | $1.2\times10^{-6}$ |
| $^{90}Sr \to ^{90}Y$ | 28.5 | 0.546 2+2.281 5 | 5.772 | 3 890 |
| $^{93}Zr \to ^{93m}Nb$ | $1.5\times10^6$ | 0.090 5 | 6.375 | 0.112 |
| $^{99}Tc$ | $2.13\times10^5$ | 0.293 6 | 6.074 | 0.710 |
| $^{107}Pb$ | $6.5\times10^6$ | 0.033 | 0.147 | 0.010 4 |
| $^{126}Sn \to ^{126m}Sb$<br>$\downarrow$ $^{126ml}Sb$<br>$\downarrow$ $^{126}Sb$ | $1\times10^5$ | 0.368+3.670 | 0.053 6 | 0.0493 |
| $^{129}I$ | $1.57\times10^7$ | 0.192 | 0.757 | 0.002 3 |
| $^{134}Cs$ | 2.062 | 2.058 5 | 0 | 8 760 |
| $^{135}Cs$ | $3.0\times10^6$ | 0.205 | 6.536 | 0.026 1 |
| $^{137}Cs \to ^{137m}Ba$ | 30.0 | 0.513 4 | 6.183 | 6 619 |
| $^{151}Sm$ | 90 | 0.076 3 | 0.419 6 | 9.34 |
| $^{154}Eu$ | 8.8 | 1.968 9 | 0 | 984 |
| $^{155}Eu$ | 4.96 | 0.252 7 | 0.032 0 | 574 |

† Only for the longer lived mother nuclide.

‡ Decay energy not particle energy (see decay schemes).

†† Thermal fssion of $^{235}U$ (fission of $^{238}U$, fssion of Pu isotopes and n, $\gamma$-reactions are important effects in a nuclear reactor).

# 20.3 Commercial Reactor Accidents

**Three Mile Island**

On March 28, 1979, the feed water pumps that supplied the steam generator of Three Mile Island Unit 2 shut down, leading to automatic shutdown of the reactor. Heat from residual radioactivity in the core caused the temperature and pressure of the reactor coolant to rise and a relief valve used to stop the repressurizer, to open. The open valve could not close and since the reactor coolant water was draining, a loss-of-coolant accident was in progress. Large amounts of radionuclides were released in the reactor building basement, but comparatively small amounts of noble gas (mainly $^{133}Xe$, 370 PBq) and $^{131}I$ (550 GBq) were released to the atmosphere and the collective effective dose is estimated to be 40 man-Sv in the surrounding area.

## Chernobyl

On April 26, 1986, a low power engineering test was being conducted at one of the reactors of the Chernobyl nuclear power station in the Ukraine (then the USSR). The reactor became unstable, resulting in thermal explosions and fires that caused severe damage to the reactor and its building. Radioactivity was released over the next ten days until the fires were extinguished and the reactor entombed in concrete. The radioactivity was released as gas and dust particles and initially blown by winds in a northerly direction. Outside Russia, the accident was first detected by increased radioactivity levels at the Forsmark nuclear power plant, about 110 km north of Stockholm, Sweden, where it caused a full alarm as the radioactivity was believed to come from the Swedish plant. Subsequently, the radioactivity released at Chernobyl was spread more to the west and southwest. For the exposed population in the Byelorussia region near Chernobyl the estimated average increased dose in the first year after the accident was approximately the same as the annual background radiation. In northern and eastern Europe in general, the increased exposure during the first year was 25%–75% above background levels. The highest dose will be delivered in southeastern Europe and is estimated to be 1.2 mSv up to year 2020, which can be compared to about 70 mSv from natural background radiation during the same period. Table 20.3 gives the fraction of the core activity released and the air and ground contamination of various nuclides at two Swedish locations.

**Table 20.3 Radionuclides released into the atmosphere in the Chernobyl accident and local contamination**

| Nuclide | Half-life | Release | | Coreinventory tot EBq | Contamination | |
|---|---|---|---|---|---|---|
| | | Tot EBq | % | | Air (Bq $m^3$) Stockholm† | Ground (kBq $m^2$) Gävle area‡ |
| $^{85}Kr$ | 10.73 y | 0.033 | 100 | 0.033 | — | — |
| $^{89}Sr$ | 50.5 d | 0.094 | 4.0 | 2.4 | — | — |
| $^{90}Sr$ | 28.6 y | 0.008 1 | 4.0 | 0.20 | — | — |
| $^{95}Zr$ | 64.0 d | 0.16 | 3.2 | 5.0 | 0.6 | 5.9 |
| $^{103}Ru$ | 39.4 d | 0.14 | 2.9 | 4.8 | 6.4 | 14.6 |
| $^{106}Ru$ | 368 d | 0.059 | 2.9 | 2.0 | 1.8 | 4.0 |
| $^{131}I$ | 8.04 d | 0.67 | 20 | 3.3 | 15 | 179 |
| $^{133}Xe$ | 5.24 d | 1.7 | ≈100 | 1.7 | — | — |
| $^{134}Cs$ | 2.07 y | 0.019 | 10 | 0.19 | 2.4 | 14.4 |
| $^{136}Cs$ | 13.2 d | — | — | — | 0.7 | 6.0 |
| $^{137}Cs$ | 30.2 y | 0.037 | 13 | 0.28 | 4.5 | 24.7 |
| $^{140}Ba$ | 12.8 d | 0.28 | 5.6 | 5.0 | 25.6 | 2.1 |

**Table 20.3**(Continued)

| Nuclide | Half-life | Release | | Coreinventory tot EBq | Contamination | |
|---|---|---|---|---|---|---|
| | | Tot EBq | % | | Air (Bq $m^3$) Stockholm† | Ground (kBq $m^2$) Gävle area‡ |
| $^{141}Ce$ | 32.5 d | 0.13 | 2.3 | 5.6 | 0.5 | 5.9 |
| $^{144}Ce$ | 284 d | 0.088 | 2.8 | 3.1 | 0.3 | 3.7 |
| $^{239}Np$ | 2.36 d | 0.97 | 3.2 | 3.0 | — | 2.7 |
| $^{238}Pu$ | 87.7 y | $3.0\times10^{-5}$ | 3 | 0.001 | — | — |
| $^{239}Pu$ | 24 100 y | $2.6\times10^{-5}$ | 3 | 0.000 9 | — | — |
| $^{240}Pu$ | 6 570 y | $3.7\times10^{-5}$ | 3 | 0.001 2 | — | — |
| $^{241}Pu$ | 14.4 y | — | 0.170 | — | — | — |

†First days, April 28-29, 1986.

‡Ullbolsta, outside of Gävle, among highest depositions outside Russia; corrected to April 28, 1986. References: USSR State Comm. on the Utilization of Atomic Energy 1986. The accident at the Chernobyl nuclear power plant and its consequences, IAEA expert meeting 25-29 Aug. 1986, Vienna.

## Fukushima

On the 11th of March 2011, an earthquake of the magnitude 9.0 on the Richter scale occurred outside the eastern coast of Japan. All affected nuclear reactors shut down as planned as a response to the earthquake. However, an unusually large tsunami hit the Fukushima Daiichi nuclear power station afterwards causing a total blackout. As the fuel heated up the zircaloy cladding underwent a highly exothermic oxidation reaction according to Eq. (20.1)

$$Zr + 2H_2O \longrightarrow ZrO_2 + 2H_2 \quad \Delta H = -6.67 + 2.57 \times 10^{-4}\ T(MJ/kg) \qquad (20.1)$$

The hydrogen produced would normally have been converted to water by the installed electrically heated recombiners or vented by the fans to the stack. However, nothing of this worked after the total power loss. Thus hydrogen escaped up into the secondary reactor building where it accumulated and later exploded. The explosion did not affect the primary containment. However, the spent fuel storage on unit 4 was damaged and the stored fuel could no longer be cooled. That increased the risk for damage not only to the fuel inside the reactor cores but also to the spent fuel in the storage pond, which was not inside an effective containment. This increased the risk for another large release of radioactive material.

The release of radioactive material from the Fukushima accident was modest when compared to Chernobyl, 5 200 PBq total from Chernobyl and about 900 PBq total into the air from Fukushima. The release at Fukushima contained little, or no, corium and consisted mainly of the volatile fission products iodine and cesium (probably in the form of CsOH, CsI, and $I_2$). However, significant land masses were contaminated. In most areas the contamination levels were comparable to those seen in many areas (e.g. close to Gävle) in Sweden after the Chernobyl accident (about 120 $kBq/m^2$) where no evacuation or cleaning up measures were deemed necessary. However in the plume north of the plant in Fukushima there were areas with about 3 MBq/m$^2$.

For several days large amounts of fission products and actinides were ejected and spread over large areas of the former USSR and Europe, see Figure 20. 1. Almost 20% of the fission products came down over Scandinavia, causing a deposition 120 kBq $^{137}$Cs/m$^2$ over the city Gävle in Sweden (170 km north of Stockholm). The plume spread down over Central Europe, causing very uneven deposition due to wind pattern and local rains, e. g., Munich, Germany, received 25 kBq $^{137}$Cs and 0. 2 kBq $^{90}$Sr per m$^2$, while Mainz (400 km away) received 180 Bq $^{137}$Cs and 0. 001 Bq $^{90}$Sr per m$^2$.

## 20. 4 Army Facility and Atomic Weapon Transport Accidents

**Kyshtym**

On September 29, 1957, a 300 m$^3$ tank containing high-level radioactive waste exploded, and about 90% (900 PBq) of the radionuclides in the tank were released and fallen in the area 300 km downwind over the next 11 h. Main released nuclides were $^{144}$Ce +$^{144}$Pr (66%), $^{95}$Zr + $^{95}$Nb (24. 9%), $^{90}$Sr + $^{90}$Y (5. 4%), $^{106}$Ru + $^{106}$Rh (3. 7%), $^{137}$Cs (0. 036%), and trace amounts of $^{89}$Sr, $^{147}$Pm $^{156}$Eu, and $^{239,240}$Pu. 1 154 people lived in an area registering 40 MBq m$^{-2}$; 1 500 people lived in an area registering 4 MBq m$^{-2}$; and 10 000 people lived in an area registering 70 kBq m$^{-2}$. Within 10 days, 1 154 people escaped from the contaminated area, and in total 10 730 people escaped from that area. The collective dose is estimated to be 2 500 man-Sv in the contaminated area.

**Windscale**

The Windscale (now called Shellafield) Reactor No. 1 was partially consumed by combustion in October 1957, resulting in the release of fission products to the surrounding countryside. The reactor was an air-cooled graphite-moderated natural-uranium reactor employed primarily for plutonium production. The radionuclides $^{131}$I (740 TBq), $^{137}$Cs (22 TBq), $^{106}$Ru (3 TBq), and $^{133}$Xe (1. 2 PBq) were released. In addition to those fission products, 8. 8 TBq of $^{210}$Po was also released, because the nuclide was produced by neutron irradiation of bismuth. The released radionuclides moved from Windscale to the south, southeast, and to London, and they contaminated vast grasslands. The collective dose is estimated to be 2 000 man-Sv in the contaminated area.

**Palomares**

Four hydrogen bombs were released from a crashed airplane on January 17, 1966, and two fell slowly with a parachute and were recovered without damage. But the remaining two bombs crashed to the earth's surface because the parachute did not open, and their normal gunpowder exploded and considerable amounts of $^{239}Pu$ and $^{240}Pu$ were scattered in an area of 2.26 $km^2$ of urban areas and uncultivated farmlands. An area of 22 000 $m^2$ was contaminated by above 1.2 MBq $m^{-2}$ of alpha-emitting nuclides. Contaminated plants and soil (to a depth of 10 cm) were collected and treated as radioactive waste. The collective dose is estimated to be 3 man-Sv in the contaminated area.

### Thule

An airplane crashed on Thule in Greenland in January 1968 and four hydrogen bombs were exploded by their gunpowder. About 10 TBq of Plutonium was recovered from the snow surface, and about 1 TBq was estimated to be taken into ice. At about 20 km from the crash point, measurable plutonium was found in the summer of the year. The exposure was not estimated because the point is near the Arctic area and people do not live there.

### SNAP-9A

A navigator satellite was launched on April 21, 1964, but failed to reach orbital speed and reentered the atmosphere at about 45 km over Indian Ocean. The satellite carried an isotopic power unit known as SNAP-9A, which contained about 6 300 GBq of $^{238}Pu$. The $^{238}Pu$ was first detected at an altitude of 30 km after 4 months, and was detected following 4 months. The plutonium decreased exponentially with the half-life of about 14 months. The sizes of the particles ranged from 5 mm to 58 mm. The $^{238}Pu$ contents of ground level-air were measured and the collective effective dose from the $^{238}Pu$ exposure in the world is estimated as 2 100 man-Sv.

### Cosmos-954

A satellite of the USSR, Cosmos-954, reentered the atmosphere over Northwest Canada on January 24, 1978. The satellite was believed to have contained 20 kg of highly enriched uranium. Released fission products were estimated since the reactor worked for 128 d at 100 kW. 75% of the particulate radionuclides were estimated to be dispersed in the atmosphere and 25% to be deposited on uninhabited regions in Northeast Canada. The vaporized radionuclides, $^{131}I$ and $^{137}Cs$, were dispersed in the atmosphere. The collective effective dose was estimated to be 20 man-Sv in the world.

## 20.5 Industrial and Medical Source Accidents

### Ciudad Juarez, Mexico

A medical source of 16.7 TBq of $^{60}Co$ was discarded by mistake in December 1983 at Ciudad Juarez in Mexico, and the 6 000 pellets of that source contaminated many products of steel, which were sold in Mexico and the USA. About 1 000 people were exposed by this source and the collective effective dose was estimated to be 150 man-Sv.

**Mochmedia, Morocco**

A $^{192}Ir$ source for the radiography of welding at construction sites was accidentally dropped in a street in September 1984, and a passenger picked it up and went home with it. All eight people of his household were exposed to 8 – 25 Sv and died, and the collective effective dose was estimated to be 80 man-Sv.

**Goiania, Brazil**

A medical source of 50. 9 TBq of $^{137}Cs$ was disposed of in September 1987 and was dismantled by a trader of scrap metal. Subsequently, 20 exposed people were hospitalized and 4 people died. For the decontamination, 42 houses were decontaminated and the total volume of 3 500 $m^3$ waste was produced and the collective effective dose was estimated to be 60 man-Sv.

**Tokaimura, Japan**

A criticality accident took place at the Tokaimura nuclear fuel processing plant in Japan on September 30, 1999, because of improper procedures. During the 24 h event, and because of only limited shielding provided by the building, some direct irradiation was measurable outside the plant site. There was only a trace release of gaseous fission products. The collective effective dose was estimated to be 150 man-Sv. Three workers inside the plant were severely exposed (5. 4 Gy of neutrons and 8. 5 Gy of gamma-rays for Person A, 2. 9 Gy of neutrons and 4. 5 Gy of gamma-rays for Person B, and 0. 81 Gy of neutrons and 1. 3 Gy of gamma-rays for Person C). Person A and B eventually died of multiple organ failure after 82 d and 210 d following the accident.

**词汇**

| | |
|---|---|
| mining | 采矿 |
| fabrication | 制造 |
| effluent | 流出物; 污水 |
| exclude | 排除 |
| geochemical | 地球化学的 |
| exhale | 呼气 |
| aquatic | 水的; 水中的 |
| collective dose | 集体剂量 |
| repressurizer | 增压器 |
| entomb | 掩埋; 封存 |
| exothermic | 放热的 |
| tsunami | 海啸 |
| recombiner | 复合器 |

evacuation 疏散
downwind 顺风
parachute 降落伞
navigator satellite 导航卫星
discard 弃置
pellet 弹丸
radiography 放射性照相法
dismantle 拆卸
organ failure 器官衰竭

**注释**

1. Table 20.1 compares the estimated collective dose to the public from the various activities of the nuclear fuel cycle, from mining and milling through nuclear waste disposal; the significant point is the recognition that the mining operation for uranium is the major contributor in the dose to the global public by quite a large fraction.

表 20.1 比较了从采矿和选矿到核废料处理的核燃料循环的各项活动对公众集体剂量的估计;重要的一点是,人们认识到,铀的开采作业是全球公众剂量的主要贡献者,占据了相当大的比例。

2. Longer lived fission products and activation products from these systems are listed as these represent the major concern to the general public if they are allowed to enter the environment; we exclude nuclides with insignificant contribution to the total activity after 10 a.

这些系统产生的寿命较长的裂变产物和活化产物之所以被列出来,是因为如果允许它们进入环境,它们会引起公众的主要关注和担忧;我们排除了 10 年后对总活度贡献不大的核素。

3. The liquid effluents from nuclear power plants and from reprocessing plants are about equally responsible for the global collective dose commitment of nuclear power generation.

核电站和后处理厂的液体流出物对全球核能发电的集体剂量承担大致同等的责任。

4. For the exposed population in the Byelorussia region near Chernobyl the estimated average increased dose in the first year after the accident was approximately the same as the annual background radiation.

对于切尔诺贝利附近白俄罗斯地区的受照射人口,事故发生后第一年估计的平均增加剂量与年本底辐射大致相同。

5. The highest dose will be delivered in southeastern Europe and is estimated to be 1.2 mSv up to year 2020, which can be compared to about 70 mSv from natural background radiation during the same period.

最高剂量将出现在东南欧,估计到 2020 年为 1.2 毫希沃特,而同期的自然本底辐射约为 70 毫希沃特。

6. In most areas the contamination levels were comparable to those seen in many areas

(e. g., close to Gävle) in Sweden after the Chernobyl accident (about 120 $kBq/m^2$) where no evacuation or cleaning up measures were deemed necessary. However in the plume north of the plant in Fukushima there were areas with about 3 $MBq/m^2$.

在大多数地区,污染水平与切尔诺贝利事故发生后瑞典许多地区(例如 Gävle 附近)的污染水平相当(约 120 $kBq/m^2$),这些地区被认为没有必要采取疏散或清理措施。然而,在福岛核电站以北的烟羽区,辐射强度约为 3 $MBq/m^2$。

7. The reactor was an air-cooled graphite-moderated natural-uranium reactor employed primarily for plutonium production.

该反应堆是一个气冷石墨慢化天然铀反应堆,主要用于生产钚。

8. 75% of the particulate radionuclides were estimated to be dispersed in the atmosphere and 25% to be deposited on uninhabited regions in Northeast Canada.

据估计,75%的微粒放射性核素分散在大气中,25%沉积在加拿大东北部无人居住的地区。

9. A medical source of 16. 7 TBq of $^{60}Co$ was discarded by mistake in December 1983 at Ciudad Juarez in Mexico, and the 6 000 pellets of that source contaminated many products of steel, which were sold in Mexico and the USA.

1983 年 12 月,墨西哥华雷斯城错误地丢弃了 16. 7 TBq $^{60}Co$ 的医疗来源,该来源的 6 000 个颗粒污染了许多在墨西哥和美国销售的钢铁产品。

# Chapter 21 Radioactive Tracer

## 21.1 Methodology

A radioactive tracer, radiotracer, or radioactive label is a chemical compound in which one or more atoms have been replaced by a radionuclide so by virtue of its radioactive decay. It can be used to explore the mechanism of chemical reactions by tracing the path that the radioisotope follows from reactants to products. Radiolabeling or radiotracing is thus the radioactive form of isotopic labeling. In biological contexts, use of radioisotope tracers is sometimes called radioisotope feeding experiments.

Radioisotopes of hydrogen, carbon, phosphorus, sulfur, and iodine have been used extensively to trace the path of biochemical reactions. A radioactive tracer can also be used to track the distribution of a substance within a natural system such as a cell or tissue, or as a flow tracer to track fluid flow. Radioactive tracers are also used to determine the location of fractures created by hydraulic fracturing in natural gas production. Radioactive tracers form the basis of a variety of imaging systems, such as, PET scans, SPECT scans and technetium scans. Radiocarbon dating uses the naturally occurring $^{14}C$ isotope as an isotopic label.

Isotopes of a chemical element differ only in the mass number. For example, the isotopes of hydrogen can be written as $^{1}H$, $^{2}H$ and $^{3}H$, with the mass number superscripted to the left. When the atomic nucleus of an isotope is unstable, compounds containing this isotope are radioactive. Tritium is an example of a radioactive isotope.

The principle behind the use of radioactive tracers is that an atom in a chemical compound is replaced by another atom, of the same chemical element. The substituting atom, however, is a radioactive isotope. This process is often called radioactive labeling. The power of the technique is due to the fact that radioactive decay is much more energetic than chemical reactions. Therefore, the radioactive isotope can be present in low concentration and its presence detected by sensitive radiation detectors such as Geiger counters and scintillation counters. George de Hevesy won the 1943 Nobel Prize for Chemistry "for his work on the use of isotopes as tracers in the study of chemical processes".

There are two main ways in which radioactive tracers are used:

(1) When a labelled chemical compound undergoes chemical reactions one or more of the products will contain the radioactive label. Analysis of what happens to the radioactive isotope provides detailed information on the mechanism of the chemical reaction.

(2) A radioactive compound is introduced into a living organism and the radio-isotope provides a means to construct an image showing the way in which that compound and its reaction products are distributed around the organism.

## 21.2 Production of Radioisotopes

The commonly used radioisotopes have short half-life and so do not occur in nature in large amounts. They are produced by nuclear reactions. One of the most important processes is absorption of a neutron by an atomic nucleus, in which the mass number of the element concerned increases by 1 for each neutron absorbed. For example,

$$^{13}C + n \longrightarrow {}^{14}C \tag{21.1}$$

In this case the atomic mass increases, but the element is unchanged. In other cases, the product nucleus is unstable and decays, typically emitting protons, electrons (beta particle) or alpha particles. When a nucleus loses a proton, the atomic number decreases by 1. For example,

$$^{32}S + n \longrightarrow {}^{32}P + p \tag{21.2}$$

Neutron irradiation is performed in a nuclear reactor. The other main method used to synthesize radioisotopes is proton bombardment. The proton is accelerated to high energy either in a cyclotron or a linear accelerator.

## 21.3 Tracer Isotopes

**Hydrogen**

Tritium ($^{3}H$) is produced by neutron irradiation of $^{6}Li$:

$$^{6}Li + n \longrightarrow {}^{4}He + {}^{3}H \tag{21.3}$$

Tritium has a half-life (4 500±8) d (approximately 12.32 a) and it decays by beta decay. The electrons produced have an average energy of 5.7 keV. Because the emitted electrons have relatively low energy, the detection efficiency by scintillation counting is rather low. However, hydrogen atoms are present in all organic compounds, so tritium is frequently used as a tracer in biochemical studies.

**Carbon**

$^{11}C$ decays by positron emission with a half-life of ca. 20 min. $^{11}C$ is one of the isotopes often used in positron emission tomography.

$^{14}C$ decays by beta decay, with a half-life of 5 730 a. It is continuously produced in the upper atmosphere of the earth, so it occurs at a trace level in the environment. However, it is not practical to use naturally-occurring $^{14}C$ for tracer studies. Instead it is made by neutron irradiation of the isotope $^{13}C$ which occurs naturally in carbon at about the 1.1% level. $^{14}C$ has been used extensively to trace the progress of organic molecules through metabolic pathways.

**Nitrogen**

$^{13}N$ decays by positron emission with a half-life of 9.97 min. It is produced by the nuclear reaction

$$^{1}H + {}^{16}O \longrightarrow {}^{13}N + {}^{4}He \tag{21.4}$$

$^{13}N$ is used in positron emission tomography (PET scan).

**Oxygen**

$^{15}O$ decays by positron emission with a half-life of 122 sec. It is used in positron emission tomography.

**Fluorine**

$^{18}F$ decays predominately by β emission, with a half-life of 109.8 min. It is made by proton bombardment of $^{18}O$ in a cyclotron or linear particle accelerator. It is an important isotope in the radiopharmaceutical industry. For example, it is used to make labelled fluorodeoxyglucose (FDG) for application in PET scans.

**Phosphorus**

$^{32}P$ is made by neutron bombardment of $^{32}S$

$$^{32}S + n \longrightarrow {}^{32}P + p \tag{21.5}$$

It decays by beta decay with a half-life of 14.29 d. It is commonly used to study protein phosphorylation by kinases in biochemistry.

$^{33}P$ is made in relatively low yield by neutron bombardment of $^{31}P$. It is also a beta-emitter, with a half-life of 25.4 d. Though more expensive than $^{32}P$, the emitted electrons are less energetic, permitting better resolution in, for example, DNA sequencing.

Both isotopes are useful for labeling nucleotides and other species that contain a phosphate group.

**Sulfur**

$^{35}S$ is made by neutron bombardment of $^{35}Cl$

$$^{35}Cl + n \longrightarrow {}^{35}S + p \tag{21.6}$$

It decays by beta-decay with a half-life of 87.51 d. It is used to label the sulfur-containing amino-acids methionine and cysteine. When a sulfur atom replaces an oxygen atom in a phosphate group on a nucleotide a thiophosphate is produced, so $^{35}S$ can also be used to trace a phosphate group.

**Technetium**

$^{99m}Tc$ is a very versatile radioisotope, and is the most commonly used radioisotope tracer in medicine. It is easy to produce in a $^{99m}Tc$ generator, by decay of $^{99}Mo$.

$$^{99}Mo \longrightarrow {}^{99m}Tc + e^- + \nu_e \tag{21.7}$$

The molybdenum isotope has a half-life of approximately 66 hours (2.75 d), so the generator has a useful life of about two weeks. Most commercial $^{99m}Tc$ generators use column chromatography, in which $^{99}Mo$ in the form of molybdate, $MoO_4^{2-}$ is adsorbed onto acid alumina ($Al_2O_3$). When the $^{99}Mo$ decays it forms pertechnetate $TcO_4^-$, which because of its single charge is less tightly bound to the alumina. Pulling normal saline solution through the column of immobilized $^{99}Mo$ elutes the soluble $^{99m}Tc$, resulting in a saline solution containing the $^{99m}Tc$ as the dissolved sodium salt of the pertechnetate. The pertechnetate is treated with a reducing agent such as $Sn^{2+}$ and a ligand. Different ligands form coordination complexes which give the technetium enhanced affinity for particular sites in the human body.

$^{99m}Tc$ decays by gamma emission, with a half-life: 6.01 hours. The short half-life ensures that the body-concentration of the radioisotope falls effectively to zero in a few days.

**Iodine**

$^{123}I$ is produced by proton irradiation of $^{124}Xe$. The caesium isotope produced is unstable and decays to $^{123}I$. The isotope is usually supplied as the iodide and hypoiodate in dilute sodium hydroxide solution, at high isotopic purity. $^{123}I$ has also been produced at Oak Ridge National Laboratories by proton bombardment of $^{123}Te$. $^{123}I$ decays by electron capture with a half-life of 13.22 hours. The emitted 159 keV gamma ray is used in single-photon emission computed tomography (SPECT). A 127 keV gamma ray is also emitted.

$^{125}I$ is frequently used in radioimmunoassays because of its relatively long half-life (59 d) and ability to be detected with high sensitivity by gamma counters.

$^{129}I$ is present in the environment as a result of the testing of nuclear weapons in the atmosphere. It was also produced in the Chernobyl and Fukushima disasters. $^{129}I$ decays with a half-life of 15.7 million years, with low-energy beta and gamma emissions. It is not used as a tracer, though its presence in living organisms, including human beings, can be characterized by measurement of the gamma rays.

**Other isotopes**

Many other isotopes have been used in specialized radiopharmacological studies. The most widely used is $^{67}Ga$ for gallium scans. $^{67}Ga$ is used because, like $^{99m}Tc$, it is a gamma-ray emitter and various ligands can be attached to the $Ga^{3+}$ ion, forming a coordination complex which may have selective affinity for particular sites in the human body.

## 21.4 Application of Radioisotopes

In metabolism research, tritium and $^{14}C$-labelled glucose are commonly used in glucose clamps to measure rates of glucose uptake, fatty acid synthesis, and other metabolic processes. While radioactive tracers are sometimes still used in human studies, stable isotope tracers such as $^{13}C$ are more commonly used in current human clamp studies. Radioactive tracers are also used to study lipoprotein mctabolism in humans and experimental animals.

In medicine, tracers are applied in a number of tests, such as $^{99m}Tc$ in autoradiography and nuclear medicine, including single-photon emission computed tomography, positron emission tomography and scintigraphy. The urea breath test for helicobacter pylori commonly used a dose of $^{14}C$ labelled urea to detect H pylori infection. If the labelled urea was metabolized by H pylori in the stomach, the patient's breath would contain labelled carbon dioxide. In recent years, the use of substances enriched in the non-radioactive isotope $^{13}C$ has become the preferred method, avoiding patient exposure to radioactivity.

In hydraulic fracturing, radioactive tracer isotopes are injected with hydraulic fracturing fluid to determine the injection profile and location of created fractures. Tracers with different half-lives are used for each stage of hydraulic fracturing. In the United States amounts per injection of radionuclide are listed in the US Nuclear Regulatory Commission (NRC) guidelines. According to the NRC, some of the most commonly used tracers include antimony-124, bromine-82, iodine-125, iodine-131, iridium-192, and scandium-46. A 2003 publication by the International Atomic Energy Agency confirms the frequent use of most of the tracers above, and says that manganese-56, sodium-24, technetium-99m, silver-110m, argon-41, and xenon-133 are also used extensively because they are easily identified and measured.

Surface properties of solids have been studied by dipping specimens into a solution containing a suitable radioactive tracer, and, after some "exposure time", removing them, washing their surface carefully and measuring the radiation emitted from them. It has been shown that a very rapid exchange takes place between atoms on a metal surface and the metal ions in solution. While the exchange is a function of the nature of the surface, within minutes it may involve atoms several hundred layers deep. The depth of penetration of sorbed radioactive isotopes can be obtained from a measurement of the absorption of the radiation or by measuring the radioactivity removed by cutting, or grinding, away thin layers. With the same technique the diffusion of atoms in their own solid matrix can be studied. For example, using single crystals of silver suspended in a solution containing silver nitrate labelled with $^{110m}Ag$, it has been possible to demonstrate different rates of diffusion into different faces of the crystal. The surface area of solids can also be determined by measurement of the sorption of radiotracers which do not penetrate into the specimen.

If a radioactive gas is incorporated in a crystalline compound the amount of gas released (the emanation ability) can be measured as a function of the temperature. It is found that the emanation increases considerably at certain temperatures, indicating structural changes in the solid at those temperatures. Studies of diffusion and emanation play a valuable role in understanding the mechanism of sintering and in the formation of new solid compounds. This has been of practical importance in the cement and glass industries, in the production of semiconductors, in the paint industry, etc. Studies of surface reactions are of practical importance for flotation, corrosion, metal plating and finishing, and detergent action to name only a few applications.

The distribution of a radioactive element or compound in a composite matrix can made visible either to the naked eye or under a microscope by means of autoradiography. The technique is based on the blackening of photographic films when exposed to nuclear radiation.

Lead is an unwanted impurity in stainless steel even in very small amounts. In order to investigate the mechanism of its incorporation, $^{212}Pb$ was added to a steel melt. After cooling, the ingot was cut by a saw and the flat surface machine-polished and etched in an electrolytic bath (electro-polishing). This provided a very flat and "virgin" surface. A photographic film was placed firmly with even pressure against the metal surface, and the film was exposed in darkness in a cool room for about a week. After development of the film, darkened spots caused by the radiation from $^{212}Pb$ showed where on the metal surface lead was present. By taking the results into account in the production process, the negative effect of lead in the raw material could be reduced.

In another technique, a polished surface of the specimen (metal, mineral, etc.) was dipped into a solution containing a radioactive reagent, which selectively reacts with one of the constituents of the surface. A mineral was dipped into potassium ethyl xanthate labelled with $^{35}S$ (be 87.5 d); the xanthate reacted selectively with sphalerite (zinc blende), ZnS, in the sample. The distribution of the xanthate, as shown by the autoradiograph, indicated the ZnS distribution in the mineral. The low b-energy of $^{35}S$, Emax 0.2 MeV, was an advantage to the technique because the resolution of the autoradiograph increases with decreasing particle range.

**词汇**

| | |
|---|---|
| tracer | 示踪剂 |
| label | 标记 |
| radioisotope | 放射性同位素 |
| cell | 细胞 |
| tissue | 组织 |
| fracture | 断裂、破裂 |
| hydraulic | 水力学 |
| imaging | 影像 |
| PET scan | 正电子发射计算机断层扫描 |
| living organism | 生命体 |
| hydrogen | 氢 |
| tritium | 氚 |
| carbon | 碳 |
| naturally-occurring | 天然形成的 |
| metabolic | 新陈代谢 |
| nitrogen | 氮 |
| positron | 正电子 |
| fluorine | 氟 |
| phosphorus | 磷 |
| biochemistry | 生物化学 |
| protein | 蛋白质 |
| neutron bombardment | 中子轰击 |
| sulfur | 硫 |
| phosphate group | 磷酸基团 |
| technetium | 锝 |
| molybdenum | 钼 |
| pertechnetate | 过硫酸盐 |
| elute | 洗脱 |
| iodine | 碘 |
| proton irradiation | 质子辐照 |
| radioimmunoassays | 放射免疫分析 |

| | |
|---|---|
| radiopharmacological | 放射药理学 |
| affinity | 亲和力 |
| glucose | 葡萄糖 |
| antimony | 锑 |
| bromine | 溴 |
| iridium | 铱 |
| scandium | 钪 |
| manganese | 锰 |
| sodium | 钠 |
| silver | 银 |
| argon | 氩 |
| xenon | 氙 |
| specimen | 样品 |
| reagent | 试剂 |
| potassium ethyl xanthate | 乙基黄原酸钾 |

**注释**

1. A radioactive tracer, radiotracer, or radioactive label is a chemical compound in which one or more atoms have been replaced by a radionuclide.

放射性示踪剂、放射示踪剂或放射性标记物是一种化合物,其中的一个或多个原子被放射性核素取代。

2. Radioisotopes of hydrogen, carbon, phosphorus, sulfur, and iodine have been used extensively to trace the path of biochemical reactions.

氢、碳、磷、硫和碘的放射性同位素已被广泛用于追踪生化反应的路径。

3. Isotopes of a chemical element differ only in the mass number.

化学元素的同位素仅在质量数上有所不同。

4. $^{129}$I is present in the environment as a result of the testing of nuclear weapons in the atmosphere. It was also produced in the Chernobyl and Fukushima disasters.

由于在大气层中进行核武器试验,$^{129}$I 被扩散到了环境之中。切尔诺贝利和福岛核事故也向环境中排放了$^{129}$I。

5. In metabolism research, tritium and $^{14}$C-labelled glucose are commonly used in glucose clamps to measure rates of glucose uptake, fatty acid synthesis, and other metabolic processes.

在新陈代谢研究中,常用氚和 $^{14}$碳标记的葡萄糖来测量葡萄糖摄取率、脂肪酸合成率和其他代谢过程。

6. In medicine, tracers are applied in a number of tests, such as $^{99m}$Tc in autoradiography and nuclear medicine, including single-photon emission computed tomography, positron emission tomography and scintigraphy.

在医学领域,示踪剂被应用于多种检测中,如 $^{99m}$Tc 用于自显影和核医学,包括单光子发射计算机断层扫描、正电子发射计算机断层扫描和闪烁扫描。

7. According to the NRC, some of the most commonly used tracers include $^{124}$Sb, $^{82}$Br, $^{125}$I,

$^{131}I$, $^{192}Ir$, and $^{46}Sc$.

根据 NRC 的说法,一些最常用的示踪剂包括 $^{124}Sb$、$^{82}Br$、$^{125}I$、$^{131}I$、$^{192}Ir$ 和 $^{46}Sc$。

8. The distribution of a radioactive element or compound in a composite matrix can made visible either to the naked eye or under a microscope by means of autoradiography.

通过自显影技术,用肉眼或显微镜都可以看到放射性元素或化合物在复合基质中的分布。

# Chapter 22 Production and Application of Radiopharmaceuticals

## 22.1 Research Reactor Production Method

As the name suggests, a radiopharmaceutical is both a pharmaceutical and radioactive. The radioactive label is used either diagnostically as an emitter of electromagnetic radiation (gamma or X-rays), of which detection allows to quantify the concentration of the radiopharmaceutical. Alternatively, the radiolabel can be used for therapeutic applications where the ionizing radiation emitted upon decay of the radionuclide is used to destroy cells. The in vivo distribution of the radionuclide in the whole body, or in specific parts of the body, can be determined by detection of gamma and/or X-rays using a single photon emission computed tomography or positron emission tomography camera. The contrast of the image will depend on the difference in concentration of the radionuclide in the target tissue relative to that in the surrounding tissue. Static imaging provides an image of the distribution of the radionuclide at a specific timepoint after tracer injection, usually when an optimal target/background ratio is obtained. Dynamic imaging acquires the concentration of the radionuclide as a function of time postinjection and can be combined with arterial sampling and quantification of radiometabolites for exact quantification of pharmacodynamic parameters such as maximum amount of receptors ($B_{max}$) and dissociation constant.

Research reactors continue to be the major source of radioisotopes production, not only for medical uses but also for industry and research, as highlighted in many presentations and several posters. An overview on the status and challenges, mostly due to aging reactors, underlined the vital roles of research reactors, with a focus on securing supplies of $^{99}$Mo/ $^{99m}$Tc and initiatives taken by the Nuclear Energy Agency Organization for Economic Cooperation & Development (NEA and OECD) through the High-Level Group on the Security of Supply of Medical Radioisotopes (HLG-MR) (involving IAEA cooperation). The emerging alternate routes for $^{99}$Mo and $^{99m}$Tc production would provide a much-needed diversity in sourcing supplies. These will supplement (but not replace) the existing commercial supplies of fission produced $^{99}$Mo and alumina column chromatography generators of $^{99m}$Tc.

The national experience of Poland in the large-scale radioisotopes production at the Maria reactor highlighted the importance of IAEA support. The radiopharmaceuticals-nuclear medicine share was reported just 1% of the global pharma industry. An IAEA-Coordinated Research Project named "Therapeutic Radiopharmaceuticals Labelled with New Emerging Radionuclides ($^{67Cu}$, $^{186}$Re, $^{47}$Sc)" (F22053) has catalysed recent developments in the production of real theranostic radioisotopes, namely $^{44/47}$Sc and $^{64/67}$Cu.

Much progress in supplementing and expanding production of $^{99}$Mo and $^{177}$Lu was reported. There were many papers on the production and use of $^{177}$Lu and the impact of $^{176}$Lu enriched target material purity on $^{177}$Lu production and resulting specific activity. This can be viewed as an indicator of the expanding production scale of carrier-added ("ca") $^{177}$Lu. There is a choice between "ca" vis-à-vis no-carrier-added ("nca") $^{177}$Lu, the latter obtained from the decay of the precursor nuclide $^{177}$Yb. It is known that $^{177}$Lu of over 740 GBq (20 Ci)/mg specific activity will suffice for most clinical applications, including those involving peptide-conjugates. At the same time, "nca" $^{177}$Lu may yet be necessary to meet specific national regulations governing the discharge limits of radioactivity in effluents in some countries. The choice of production option for $^{177}$Lu of required specific activity will be dictated by ease of access to and cost of (i) $^{176}$Lu enriched target and very high neutron flux; (ii) $^{176}$Yb enriched target and radiochemical processing for separation of $^{177}$Lu from Yb.

## 22.2 Accelerator Production Method

Charged particle accelerators, proton cyclotrons in particular — that include compact MCs — are extensively deployed for regular production of a variety of radioisotopes, especially positron emitters. An overview of the field was presented, highlighting the production routes of the less-commonly used (so called "non-standard") positron emitters, e. g. $^{86}$Y, $^{64}$Cu, and $^{89}$Zr, using low-energy accelerators, and of therapeutic radioisotope, e. g. $^{67}$Cu, $^{225}$Ac, $^{47}$Sc, $^{193m}$Pt, and $^{117m}$Sn, in medium-energy accelerators.

The production of radionuclides of metallic elements-viz. $^{89}$Zr, $^{43/47}$Sc, $^{52}$Mn, and $^{45}$Ti, in the TR-24 MCs at University of Alabama at Birmingham (UAB), USA, is based on unique procedures adopted in the design and preparation of solid pellet targets and separation methods. UAB's MC of Ep 18 MeV is used for the production of $^{89}$Zr and $^{45}$Ti, using mono-nuclidic targets in both cases and Al foil for energy degradation.

The highlight of the session was the progress and notable achievements in the production of radiometals using liquid targets, especially of $^{68}$Ga from enriched $^{68}$Zn target in MC. The experience gained in Portugal in tens of GBq (curie level) production of GMP grade $^{68}$Ga, at very low cost too, is significant. This signals the end of the monopoly of $^{68}$Ge-$^{68}$Ga generator suppliers.

The use of $^{68}$Ga for PET/CT has rapidly grown over the world and its use is now second only to $^{18}$F. Hence, the $^{68}$Zn(p, n) route of production in MC (Ep 13–14 MeV) is becoming well-established (with IAEA support in technology evaluation through a coordinated research project called "Production of cyclotron-based Gallium-68 radioisotope and related radiopharmaceuticals") and is acknowledged as an important advancement; e. g., in Canada, yields of 33.3 GBq (0.9 Ci) from 6 mm target and 144.3 GBq (3.9 Ci) from 10 mm target (shelf-life 5 h) were achieved. Both solid target and liquid target are developed and deployed for Ci level production of $^{68}$Ga. In the USA, for GBq (Ci) level production, a target design and processing method are being patented.

Shipments of $^{68}$Ga to nuclear medicine sites around medical cyclotrons (MC) centres are nowadays routine, similar to that of $^{18}$F products. Analysing the pros and cons of this development is helpful. Ease of access to $^{68}Ga$ using MC and ending monopoly of sourcing of $^{68}Ga$ are attractive features. On the other hand, the advantage to generator sources of ready access to $^{68}Ga$, even without MC on-site, becomes lost. However, $^{68}Ge$ production is limited to a few centres and it is not readily commercially available for users other than specific contractual procurers, namely, industrial manufacturers of $^{68}Ge$-$^{68}Ga$ generators.

Interest in the use of longer-lived positron emitters to match certain physiologic parameters and slow delivery vector molecules is growing. $^{89}Zr$ and $^{45}Ti$ are under active consideration in this context, and presentations of such included at the ISTR-2019. These radioisotopes are produced using mononuclidic target elements, Y and Sc, respectively (no enriched target required). At the University of Alabama at Birmingham, 18 MeV MC has been used with energy degradation Al foils; typical yields of radioisotope reported are:

— $^{89}Y(p, n)^{89}Zr$, Ep 13 MeV, 3 h, 40 μA, 2.6 GBq (70 mCi);

— $^{45}Sc(p, n)^{45}Ti$, Ep 13–14MeV, 0.5 h, 10 μA, 2.3 GBq (63 mCi).

The scope of radiometals production in a 30 MeV cyclotron was reported (RFT-30 machine in RoK), with the focus on $^{89}Zr$, $^{68}Ge$, $^{44}Sc$, and $^{67}Cu$, along with respective target details and processing methods.

Deferoxime (DFO) is a popular chelator molecule in use with $^{89}Zr$; for example, $^{89}Zr$-DFO-Trastuzumab has been used in breast cancer cases for immuno-PET imaging. Research and development (R&D) on other chelators are also prevalent. PSMA-11 inhibitor-binder ligand used with $^{68}Ga$ has also been used with $^{45}Ti$ for prostate cancer metastases in PET imaging.

The presentation of experience at the Canadian TRIUMF accelerator facility on the production and application of $^{225}Ac$ and/or $^{213}Bi$ included an overview of the targeted alpha therapy concept as well as an outline for future plans.

## 22.3 Research Reactor Production Method

IAEA activities to foster the sustainable supply of $^{99}Mo$ remain important and attractive for many member states; there is room for more participants to contribute. The session reminded participants of safeguards related aspects due to the use of enriched $^{235}U$ targets. IAEA has run coordinated research projects to evaluate options for production of $^{99}Mo$ and $^{99m}Tc$ in accelerators. They required very highly enriched $^{100}Mo$ targets to meet the radionuclidic purity specifications.

The largest $^{99m}Tc$ consumer country, USA, has taken up several measures for domestically securing $^{99}Mo$ supplies. In particular, new ways of producing $^{99}Mo$ and $^{99}Mo$-$^{99m}Tc$ generators are being successfully developed in the USA. New initiatives such as accelerator-based production of $^{99}Mo$ opened up new opportunities. Progress by several companies offer new solutions but mostly limited to the USA.

Other generator systems were discussed in a presentation from Troitsk, Russia, mainly with the perspective of producing the parent radionuclides by using large facilities, such as the high energy proton accelerator at the Institute for Nuclear Research in Troitsk, Russia. The availability of $^{82}Sr$-$^{82}Rb$ generators has significantly increased and several clinical trials were initiated. $^{82}Rb$ is in considerable demand for myocardial perfusion PET imaging, especially in the aftermath of the 2008–2010 crisis in $^{99}Mo/^{99m}Tc$ supplies. $^{225}Ac$ can be produced in Troitsk, but with 0.2% of $^{227}Ac$ impurity. Hence, this can rather be used for the $^{225}Ac$-$^{213}Bi$ generators instead of direct use of $^{225}Ac$ in patients.

A quasi-generator system utilizing accelerator produced $^{47}Ca$ as a parent for $^{47}Sc$ was presented by the Institute of Nuclear Chemistry and Technology, Poland. However, the small difference in half-life of $^{47}Ca$ (4.5 d) and $^{47}Sc$ (3.35 d) and the very low natural abundance of $^{46}Ca$ (0.004%) would be adverse factors in this approach.

## 22.4 Application of Radiopharmaceuticals

### Theranostic

The growing interest in thetheranostic approach, that is, using the same vector molecule targeted to the disease lesion for delivering the diagnostic radioisotope for imaging (mostly by PET) and therapeutic radioisotope (usually beta emitters and alpha emitters) for targeted therapy, is driving considerable R&D efforts in many countries.

Interest in the potential strength of mAb (despite slow kinetics) for highly selective and specific targeted imaging is reviving due to using longer lived $^{89}Zr$ (78.4 h) as a nearly ideal radioisotope for labelling mAb. Examples of $^{89}Zr$ labelled antibodies for cancer imaging are:

(1) $^{89}Zr$-DFO-Trastuzumab for imaging HER2 + tumours or for following-up response to inhibitor therapy of gastric cancer; and

(2) $^{89}Zr$-DFO-HuMAb-5B1 for targeting the carbohydrate antigen CA19.1 over-expressed by pancreatic tumours.

To address the slow kinetics with antibodies, pre-targeting is also pursued as a solution, e.g., using tetrazine for in-vivo click chemistry, and $^{64}Cu$ / $^{18}F$ for labelling for PET.

Rich experience and expertise are now available in many centres in the production of $^{68}Ga$-DOTATATE and $^{177}Lu$-DOTATATE for treatment of neuroendocrine tumours (NET). There is a choice of "manual-processing", "automated module" or "kit-based" procedures available.

Radiopharmaceuticals for targeting PSMA in prostate cancer patients by using paired radioisotope, $^{68}Ga$ (for PET imaging) and $^{177}Lu$ (for therapy), are of very high interest currently. Recently, targeting PSMA with radiotherapeutic agents labelled with alpha emitter $^{225}Ac$ is emerging as a superior alternative to $^{177}Lu$.

Focus on the "isotope on line separation device" (ISOLDE) technology—a strategy for isolating isotopes generated in situ, based on their mass difference and applying the technique of "laser resonance ionization" forevaporating nuclides from an irradiated sample—is essential when dealing with the production of unconventional (exotic) radioisotopes in an isotopic mixture. In other words, this is an attempt to enrich isotopes of elements in situ, as they are being generated by spallation reaction or similar. Such innovative sourcing of radioisotopes is however unlikely to make an impact on clinical practices in the near future.

$^{64}Cu$ (12.7 h, 17% positron emission) is envisaged to emerge as a theranostic nuclide (akin $^{131}I$), as it can be used for both imaging and for therapy (beta emitter). Also, $^{67}Cu$ can be used for therapy, keeping $^{64}Cu$ for positron emission tomography imaging alone. $^{64}Cu$ can be produced preferably in medical cyclotrons using enriched $^{64}Ni$ target [$^{64}Ni(p, n)$] and in research reactors using enriched $^{64}Zn$ target [$^{64}Zn(n, p)$] with lower yields. An IAEA-coordinated research project named "Copper-64 Radiopharmaceuticals for Theranostic Applications", has been dedicated to the production and application of $^{64}Cu$-theranostic Radiopharmaceuticals since 2016. Furthermore, copper is involved in human biochemical processes so that interest in $^{64}Cu$ based radiopharmaceuticals has been accordingly quite high, although until recently no major impact has been made in nuclear medicine practices. $^{64}Cu$ in a simple divalent inorganic form, $CuCl_2$, has been recently shown to be useful for imaging many tumours. Several other $^{64}Cu$ products are under investigation. For example, production and radiolabelling of an antibody against the class of glycoproteins "mucins", expressed by breast cancer cells — radio-conjugate $^{64}Cu$-NOTA-Anti-MUC1 — has been prepared and were characterized. Pre-clinical results were done in animal models and were presented by a research group from Iran.

**Single photon emission computed tomography (SPECT)**

This session emphasises the continuing importance, relevance and value of $^{99m}Tc$ Radiopharmaceuticals, especially in light of advances in imaging instrumentation. For example, the launch of an ultra-high resolution, fast acquisition system, G-SPECT, by a European company, as well as advances in chemistry approaches for linking Tc with biologically sound vector molecules; e.g., use of macrocyclic ligand Nodaga as chelator for Tc and Re; tricarbonyl chemistry aided conjugation to antagonist peptides to use with $^{99m}Tc$ and $^{188}Re$. There is merit to revisit certain $^{99m}Tc$ Radiopharmaceuticals to perform clinical trials again, e.g., $^{99m}Tc$-Teboroxime for myocardial perfusion imaging, as well as support resurgence of R&D. Consequently, a natural poser will be to use $^{123}I$ products and superior SPECT systems. Now that over 1 300 MC centres are in operation across the world (IAEA cyclotron database, presented in Session 14) and the capacity to distribute short-lived radioisotope is available, revisiting $^{123}I$ (13 h) based Radio-pharmaceuticals may be warranted.

Based on the success of PSMA targeting agents and PET imaging in prostate cancer patients, development of a SPECT agent, $^{99m}Tc$ based binder to PSMA — called HYNIC-iPSMA — is of interest in Mexico. It has also been used with therapeutic nuclides $^{177}Lu$ and $^{225}Ac$ by using another linker DOTA (M-DOTA-HYNIC-iPSMA). Another reported use was the synthesis and biological evaluation of $^{99m}Tc$ based integrin targeting agents by using RGD peptide linker conjugate.

The session had two general presentations, one covering many of the issues faced by developing countries in providing/availing expert nuclear medicine and radiopharmaceutical resources, and the other on the rich 60 a experience with indigenous radiopharmaceutical programme in Brazil and its future plans for expansion.

**Positron emission tomography (PET)**

This session presented glimpses of some of the advances and trends involving mainly products of $^{18}F$, $^{68}Ga$ and $^{124}I$. Apart from several $^{18}F$ labelled organic molecules under development and/or in regular use and continual development of $^{18}F$ radiolabelled NOTA/NODAGA peptides via Al $^{18}F$, the success in targeting PSMA in prostate cancer patients with $^{68}Ga$ labelled binder molecules has led to researchers exploring $^{18}F$ based binder molecules to target PSMA. Some novel synthesis of $^{18}F$-binder-PSMA via click chemistry was presented by an expert from the Netherlands. In another presentation a researcher from Thailand reported that $^{18}F$-PSMA-1007, with a shelf life of 8 h, allowed PET imaging of 8 patients as compared to only 3 with $^{68}Ga$-PSMA-11.

Cyclotron production of $^{124}I$ for PSMA binder radiolabelling was reported by Saudi Arabia, with high radiolabelling yields, very high purity and specific activity.

With growing use of $^{68}Ga$ peptide conjugates, automated formulation methods for $^{68}Ga$-DOTATOC and DOTATATE developed in Innsbruck, Austria, were described. The use of SSTR antagonist ligands for targeting neuroendocrine tumours and of $^{68}Ga$ PSMA binder conjugate for molecular diagnosis of prostate cancer was also cited. Fibroblast activation protein linked to $^{68}Ga$ ($^{68}Ga$-FAPI) has been used to diagnose more than 20 human cancers.

$^{64}Cu$ labelled MeCOSar-Tyr3-octreotate ($^{64}Cu$-SARTATE) — a new product developed due to the success of $^{68}Ga$-DOTA-Octreotate or $^{68}Ga$-DOTATATE — was evaluated for PET imaging in Australia for the localization of SSTR-positive NETs and dosimetry planning for personalized peptide receptor radionuclide therapy (PRRT, PRRNT). The rationale was to take advantage of the longer half-life of $^{64}Cu$ (12. 7 h), compared to that of $^{68}Ga$ (68 min), to better match with bio-kinetics and personalised dosimetry for therapy.

**Therapy**

This session attempted to cover Radiopharmaceuticals for both the established metastatic bone pain palliation applications and the targeted therapy for tumours.

The success of using $^{89}Sr$ for bone pain palliation in terminal cancer patients could not be expanded to many parts of the world due to the need to import the product at high cost. At the University of Missouri, Columbia, USA, scientists investigated the use of phosphonate conjugated radiometal ($^{153}Sm$) to enhance uptake at tumour met sites to mitigate pain. The resultant, FDA approved $^{153}Sm$-EDTMP has been used widely since the late 1990s, as $^{153}Sm$ can be more easily produced than $^{89}Sr$. Thanks to the IAEA coordinated research project support, the analogous product $^{177}Lu$-EDTMP has been developed and launched about ten years ago, where clinical trials were also facilitated. Its usage has been expectedly on the rise, as the longer half-life of $^{177}Lu$ helps shipping products to nuclear medicine centres in even remote locations. Furthermore, the lower range beta emitters ($^{153}Sm$, $^{177}Lu$) offer greater advantages for bone pain palliation in terms of lower bone marrow toxicity and, in turn, enables delivery of higher dose to tumours. In case of preferred prolonged delivery of radiation dose to bone mets, $^{170}Tm$-EDTMP ($^{170}Tm$, 120 d) can be considered. Mixed radioisotope therapy approach is also possible.

The emergence of the use of $^{223}Ra$ as an alpha emitter and strontium analogue is an important addition to the session topic. The initial excitement of superior longevity in patients treated with $^{223}Ra$ could not be realised when applied in larger groups of patients. Selection of patients is a crucial factor for both alpha therapy and beta therapy.

The successful use of peptide based targeted therapy of neuro-endocrine tumour (NET) mets and of enzyme inhibitor ligand based targeted therapy for prostate cancer mets, both using $^{177}Lu$, has also led to harnessing high-linear energy transfer alpha emitter or Auger electron emitter radioisotopes. The most widely used therapeutic counterpart for $^{68}Ga$ remains $^{177}Lu$, while recent advances in radionuclide production methods have made $^{225}Ac$, $^{213}Bi$ and $^{161}Tb$ available as alternatives to $^{177}Lu$. This has created interesting opportunities to treat mets with a short-range alpha and/or Auger and conversion electron emissions.

A later section of the report is devoted to alpha emitters, while certain subtle aspects covered in the current session are outlined here. This refers to the challenge with (multiple) alpha emitters like $^{225}Ac$ (10 d) undergoing a series of decay with the (likely) release of its daughter nuclide(s) from the chelator; this can result in normal tissue toxicity. For example, Francium (Fr) is a potassium analogue, and $^{221}Fr$ (4.9 min), decay product of $^{225}Ac$, can be taken up in normal tissues including myocardium. Consequently, an option to use separated $^{213}Bi$ (46 min), in place of $^{225}Ac$, has been proposed by some researchers. $^{213}Bi$ is a mono alpha emitter, but its kidney burden is higher; hence, pre-targeting type approach may be a solution for its future use. The use of Auger electron emitter $^{161}Tb$ resulted in more tumour regression and longer survival, compared to that of $^{177}Lu$.

Other presentations in this session described three methods and products, two from India and one from Malaysia. Development of a robust kit method to prepare $^{188}ReN$-DEDC/lipiodol in a hospital radiopharmacy, free of glacial acetic acid (glacial acetic acid can result in failures of synthesis, if volume is not tightly controlled) was reported from India. Oxalic acid was used in this method to eliminate the use of glacial acetic acid. This method allowed variation in the radioactive solution volume up to 5 mL (cf. 3 mL with traditional preparation) and faster kinetics of reaction — within 5 minutes compared to 30 minutes with glacial acetic acid. This study was meant to develop import substitution, as an alternative to labelled microspheres or use of HDD/lipiodol with various radionuclides.

$^{90}Y$-acetate obtained from high level liquid waste from a two-stage supported liquid membrane (SLM)-based $^{90}Sr/^{90}Y$ generator of India was used to prepare $^{90}Y$-DOTATATE and evaluated in SSTR2 positive tumour cells, in comparison to $^{68}Ga$-DOTATATE and $^{177}Lu$-DOTATATE. Comparable radiolabelling and uptake were observed.

In Malaysia, polystyrene microparticles containing $^{152}Sm$ were prepared and irradiated to produce $^{153}Sm$ labelled product for evaluation for hepatic embolization, as an alternative to the imported $^{90}Y$ microspheres. No degradation upon irradiation and no radioisotope impurities were found in the study.

## 词汇

radiopharmaceutical 放射性药物
pharmaceutical 制药的
single photon emission computed tomography 单光子发射计算机断层成像术
target tissue 靶组织

| | |
|---|---|
| arterial sampling | 动脉取样 |
| alumina column chromatography generators | 氧化铝柱层析 |
| pharma industry | 制药行业 |
| therapeutic | 治疗的 |
| catalyze | 催化 |
| theranostic | 治疗诊断学 |
| carrier-added | 携带载体 |
| vis-à-vis | 相对于 |
| precursor nuclide | 前驱核素 |
| clinical application | 临床应用 |
| peptide-conjugate | 肽偶联物 |
| positron emitter | 阳电子发射器 |
| energy degradation | 能量降解 |
| chelators | 螯合剂 |
| myocardial perfusion | 心肌灌注 |
| quasi-generator | 准发生器 |
| Trastuzumab | 曲妥单抗 |
| carbohydrate antigen | 糖链抗原 |
| tetrazine | 四嗪 |
| neuroendocrine tumours | 神经内分泌肿瘤 |
| situ | 原位 |
| laser resonance ionization | 激光共振电离 |
| spallation reaction | 散裂反应 |
| radiopharmacy | 放射性药物学 |
| integrin | 整合蛋白 |
| peptide | 肽 |
| antagonist ligand | 拮抗剂配体 |
| fibroblast | 纤维原细胞 |
| dosimetry | 放射量测定 |
| metastatic | 转移性的 |
| marrow toxicity | 骨髓毒性 |
| radiation dose | 辐射剂量 |
| enzyme inhibitor | 酶抑制剂 |
| lipiodol | 碘化油 |
| acetate | 醋酸盐 |
| hepatic embolization | 肝栓塞 |

**注释**

1. Dynamic imaging acquires the concentration of the radionuclide as a function of time postinjection and can be combined with arterial sampling and quantification of radiometabolites for

exact quantification of pharmacodynamic parameters such as maximum amount of receptors (Bmax) and dissociation constant.

动态成像获取放射性核素的浓度作为注射后时间的函数,并可以与动脉取样和放射性代谢物的定量相结合,以精确地定量药效学参数,如最大受体量($B_{max}$)和解离常数。

2. The radiopharmaceuticals-nuclear medicine share was reported just 1% of the global pharma industry. An IAEA-Coordinated Research Project named "Therapeutic Radiopharmaceuticals Labelled with New Emerging Radionuclides ($^{67}$Cu, $^{186}$Re, $^{47}$Sc)" (F22053) has catalysed recent developments in the production of real theranostic radioisotopes, namely $^{44/47}$Sc and $^{64/67}$Cu.

据报告,放射性药物-核医学仅占全球制药业的1%。国际原子能机构协调的一个名为"用新出现的放射性核素($^{67}$Cu、$^{186}$Re、$^{47}$Sc)标记的治疗性放射性药物"(F22053)的研究项目,推动了真正的治疗性放射性同位素(即$^{44/47}$Sc 和$^{64/67}$Cu)生产的最新发展。

3. $^{82}$Rb is in considerable demand for myocardial perfusion PET imaging, especially in the aftermath of the 2008-2010 crisis in $^{99}$Mo/$^{99m}$Tc supplies.

心肌灌注 PET 成像对$^{82}$Rb 的需求量很大,尤其是在 2008—2010 年 $^{99}$Mo/$^{99m}$Tc 出现供应危机之后。

4. The growing interest in the theranostic approach, that is, using the same vector molecule targeted to the disease lesion for delivering the diagnostic radioisotope for imaging (mostly by PET) and therapeutic radioisotope (usually beta emitters and alpha emitters) for targeted therapy, is driving considerable R&D efforts in many countries.

人们对治疗方法的兴趣与日俱增,这种方法是指使用针对疾病病灶的同一载体分子,来提供用于成像的诊断性放射性同位素(主要通过正电子发射计算机断层扫描)和用于靶向治疗的治疗性放射性同位素(通常是 β 发射体和 α 发射体),这将推动许多国家开展大量研发工作。

5. $^{64}$Cu labelled MeCOSar-Tyr3-octreotate ($^{64}$Cu-SARTATE) — a new product developed due to the success of $^{68}$Ga-DOTA-Octreotate or $^{68}$Ga-DOTATATE — was evaluated for PET imaging in Australia for the localization of SSTR-positive NETs and dosimetry planning for personalized peptide receptor radionuclide therapy (PRRT, PRRNT).

$^{64}$Cu 标记的 MeCOSar-Tyr3-octreotate($^{64}$Cu-SARTATE)是在$^{68}$Ga-DOTA-octreotate 或$^{68}$Ga-DOTATATE 取得成功的基础上开发的新产品,在澳大利亚进行了 PET 成像评估,用于 SSTR 阳性 NET 的定位和个性化肽受体放射性核素治疗( PRRT,PRRNT)的剂量规划。

6. The successful use of peptide based targeted therapy (PRRT) of neuro-endocrine tumour (NET) mets and of enzyme inhibitor ligand based targeted therapy for prostate cancer mets, both using $^{177}$Lu, has also led to harnessing high-linear energy transfer alpha emitter or Auger electron emitter radioisotopes.

基于肽的神经内分泌肿瘤转移靶向治疗和基于酶抑制剂配体的前列腺癌转移靶向治疗(均使用$^{177}$Lu)的成功应用,也促使人们开始利用高线性能量转移 α 发射体或俄歇电子发射体放射性同位素。

7. $^{90}$Y-acetate obtained from high level liquid waste from a two-stage supported liquid membrane (SLM)-based $^{90}$Sr/$^{90}$Y generator of India was used to prepare $^{90}$Y-DOTATATE and evaluated in SSTR2 positive tumour cells, in comparison to $^{68}$Ga-DOTATATE and $^{177}$Lu-

DOTATATE. Comparable radiolabelling and uptake were observed.

$^{90}$Y-acetate 取自印度一个基于两级支撑液膜(SLM)的$^{90}$Sr/$^{90}$Y 发生器产生的高浓度液体废物,用于制备 $^{90}$Y-DOTATATE 并在 SSTR2 阳性肿瘤细胞中进行评估,并与 $^{68}$Ga-DOTATATE 和 $^{177}$Lu-DOTATATE 进行比较,观察到了相似的放射性标记和摄取。

# Chapter 23 Complex Inorganic Actinide Materials

Actinides can complex in a wide variety of ways with fluorides, borates, sulfates, and phosphates, which together encompass a large number of known actinide structures. The topography of the present structures can commonly depend on a wide variety of factors such as the oxidation state, ionic radii, stoichiometry, pH, and temperature. Most of the synthetic reactions are conducted in either the solid state or hydrothermally — in general, this also mimics the expectations for chemistry found in either reactor conditions or in a repository environment. To study the topography and resulting structures in detail, single-crystal x-ray diffraction is considered to be the most robust and reliable technique at hand. Nevertheless, a large number of experimental techniques can be employed to discover structural information, and this will be presented as each individual case studied permits.

## 23. 1 Fluorides

**Trivalent and tetravalent fluorides**

The beginning of structural actinide chemistry was ushered in by the prolific crystallographer and Manhattan project scientist William Zachariasen. This beginning saw many improvements in the interpretation of crystallographic information, such as weak diffraction lines and absorption corrections. The need to characterize fluoride complexes first came from the ubiquitous use of $UF_6$ in gas centrifuges for the isotopic enrichment of $^{235}U$ from $^{238}U$. While none of these compounds are necessarily the most complex, they are important to detail as the basic building blocks to higher-order structures.

Crystal parameters for the trivalent and tetravalent fluorides are listed in Table 23. 1. The trivalent fluorides crystallize predominantly in the hexagonal crystal system and are isostructural with the $LaF_3$ structure type. They exhibit the very common behavior of a decrease in overall size due to the decrease in ionic radii across the series.

Depending on the actinide, different treatments to achieve the trivalent fluorides are necessary. For example, since curium in solution already resides preferably in the trivalent state, it can be precipitated out by simply adding HF. However, for an element like plutonium, it is necessary to maintain a reducing environment with the use of hydroxylamine or ascorbic acid with hydroxylamine.

The tetravalent fluorides crystallize in the monoclinic crystal system and are isostructural with the $ZrF_4$ structure type. $UF_4$, as a group example, has two distinct uranium sites which are each arranged in a slightly distorted antiprism configuration. There are 12 repeating formula units per unit cell. Curium, which resists oxidation to the tetravalent state, is reacted with purified fluorine gas at 400 ℃.

Most of the trifluorides shown in Table 23. 1 exhibit the $LaF_3$ structure type. However, it should be noted that both berkelium and californium have two entries in this table for the trifluoride. This is because for each of these, the structure type is related to a function of temperature. While both do adopt the $LaF_3$ structure type at higher temperatures, they assume the $YF_3$ structure type in the orthorhombic crystal system at 25 ℃.

**Table 23. 1 Colors and lattice parameters for some trivalent and tetravalent fluorides**

| Compound | Color | $a_0$(Å) | $b_0$(Å) | $c_0$(Å) | β(°) |
|---|---|---|---|---|---|
| $AcF_3$ | White | 7. 41 | 7. 41 | 7. 55 | 90 |
| $ThF_4$ | White | 13. 10 | 11. 01 | 8. 6 | 126 |
| $PaF_4$ | Brick-red | 12. 86 | 10. 88 | 8. 54 | 126. 34 |
| $UF_3$ | Black | 7. 181 | 7. 181 | 7. 348 | 90 |
| $UF_4$ | Blue-green | 12. 803 | 10. 792 | 8. 372 | 126. 30 |
| $NpF_3$ | Purple | 7. 129 | 7. 129 | 7. 288 | 90 |
| $NpF_4$ | Green | 12. 67 | 10. 62 | 8. 31 | 126. 16 |
| $PuF_3$ | Purple | 7. 092 | 7. 092 | 7. 240 | 90 |
| $PuF_4$ | Dusty-pink | 12. 59 | 10. 55 | 8. 26 | 126. 16 |
| $AmF_3$ | Pink | 7. 044 | 7. 044 | 7. 225 | 90 |
| $AmF_4$ | Rose-tan | 12. 49 | 10. 47 | 8. 19 | 126. 16 |
| $CmF_3$ | White | 6. 999 | 6. 999 | 7. 129 | 90 |
| $CmF_4$ | Yellow-green | 12. 45 | 10. 45 | 8. 16 | 126 |
| $BkF_3$ | Yellow-green | 6. 70 | 7. 09 | 4. 41 | 90 |
| $BkF_4$ | Yellow-green | 6. 97 | 6. 97 | 7. 14 | 90 |
| $BkF_3$ | Yellow-green | 12. 396 | 10. 466 | 8. 118 | 126. 33 |
| $CfF_3$ | Green | 6. 653 | 7. 039 | 4. 393 | 90 |
| $CfF_3$ | Green | 6. 945 | 6. 945 | 7. 101 | 90 |
| $CfF_4$ | Green | 12. 327 | 10. 402 | 8. 113 | 126. 44 |

The number of complex fluorides is vast. A simple step beyond the tetravalent fluorides, for example, is to complex alkali fluorides together with the actinides in order to create a ternary tetravalent fluoride. It just happens that this is very relevant due to the widespread possibility of developing molten salt reactor designs. These reactors can use either eutectic mixtures of either LiF or NaF as the solvent to maintain fuel and fission products for reprocessing. Higher temperatures are possible with these designs, which can allow for more efficient burn-up. However, depending upon the reactor conditions, a spectrum of redox potentials makes for potentially diverse chemical environments. Therefore, various conditions for forming these complexes have been of interest. With the element protactinium, it can be seen that reducing conditions at 400 ℃ can make tetravalent complexation possible.

**Pentavalent and hexavalent fluorides**

Undoubtedly, the higher oxidation states are represented by a smaller margin of the actinide elements. Nevertheless, because a strong suit of these compounds is represented by uranium and plutonium, their complexes with fluoride are important in a range of applications. Besides $UF_6$ and $PuF_6$, $NpF_6$ is possible and shares similar properties as its two neighbors on the periodic table. All of the hexavalent fluorides exist in the tetragonal crystal system and have low-boiling temperatures. Meanwhile, the pentavalent class includes protactinium but loses plutonium. This slight shift to the left of the 5f-series also exhibits a significantly lower stability of $NpF_5$ which disproportionates to $NpF_4$ and $NpF_6$ under a majority of circumstances. $NpF_5$, as do all of the binary pentavalent fluorides, crystallizes in the orthorhombic crystal system. All of the binary species are listed in Table 23. 2.

**Table 23. 2 Colors and lattice parameters for some pentavalent and hexavalent fluorides**

| Compound | Color | $a_0$(Å) | $b_0$(Å) | $c_0$(Å) |
|---|---|---|---|---|
| $PaF_5$ | White | 11. 530 | 11. 530 | 5. 190 |
| α-$UF_5$ | Bluish-white | 6. 525 | 6. 525 | 4. 472 |
| β-$UF_5$ | Greenish-white | 11. 473 | 11. 473 | 5. 209 |
| $NpF_5$ | Bluish-white | 6. 536 | 6. 536 | 4. 456 |
| $UF_6$ | White | 9. 900 | 8. 962 | 5. 207 |
| $NpF_6$ | Orange | 9. 910 | 8. 970 | 5. 210 |
| $PuF_6$ | Reddish brown | 9. 950 | 9. 020 | 5. 260 |

Ammonium fluoride, as well as all of the alkali metals, also forms ternary complexes with pentavalent fluorides. Stoichiometric ratios of 1:1, 2:1, and 3:1 provide, in most cases, distinct products resembling the formulas $AAnF_6$, $A_2AnF_7$, and $A_3AnF_8$, where A = $NH_4$, Na, K, Rb, and Cs. Li, regardless of stoichiometric ratio, favors the 1:1 ratio. These reactions only generally apply to the pentavalent actinides listed in Table 23. 2, though pentavalent plutonium can also be stabilized by a number of these reactions. Another important observation is that if greater stoichiometric ratios are used, the products remain at the 3:1 ratio. This is unlike the tetravalent fluoride complexes, where several stable and unique compounds can exist with the synthetic use of the 4:1 ratio. For uranium, this tendency can be referred to commonly as the F/U limit. The F/U limit of 8 is the maximum ratio of fluorine atoms that can structurally coexist per uranium atom in a ternary alkali uranium fluoride complex regardless of the valence state. Therefore, the expectation for complex hexavalent uranium fluorides would be a synthetic limit of the stoichiometric ratio held at 2∶1. Such examples include $CsUF_7$, $NH_4UF_7$, and $Na_2UF_8$, with other examples seen elsewhere in the literature.

## 23.2 Borates

Actinide borates are considered to be one of the most successful oxyanions for probing structural variation across the 5f-series. This is because borates, most commonly derived from boric acid, can adopt many structure types depending upon only slight changes in reaction conditions. Examples of such conditions include pH, temperature, cation size, stoichiometry, and counterions. A large number of complex topologies exist. Furthermore, borates have become synonymous with the growing evidence that indicates increased participation in bonding by 6d and 5f orbitals as the series is traversed.

There are several different synthetic strategies for making uranyl borates. In the first strategy, high-temperature melts with $B_2O_3$ result in structures dominated by $UO_6$ and $UO_7$ polyhedra. These reactions were commonly run in the 1980s and exhibited coordination with the borate in the form of $BO_3$ triangles. One example of this is the non-centrosymmetric complex $Ca(UO_2)_2B_2O_6$. In the second strategy, slow evaporation with mixtures of $UO_2^{2+}$ and borate at room temperature result in the maintained uranyl core surrounded by isolated clusters of cyclic polyborates. This example was the first uranium borate ever made and was determined to be $K_6[UO_2(B_{16}O_{24}(OH)_8)] \cdot 12H_2O$. Finally, the most recently developed boric acid flux reactions are also considered to be an effective way to make borates. Heated with a thorium salt at 200 ℃, an inorganic thorium borate complex results:

$$Th(NO_3)_4 \cdot 5H_2O + 6H_3BO_3 \longrightarrow [ThB_5O_6(OH)_6][BO(OH)_2] \cdot 2.5H_2O + 4HNO_3 + 5.5H_2O \tag{23.1}$$

Named NDTB-1, this material crystallizes in the cubic space group Fd-3. One of the special attributes of these borates can be seen right away with the ability of separate $BO_4$ tetrahedra and $BO_3$ triangles to form within the lattice. This type of borate dichotomy has a multiplicative effect on the potential types of structural topographies that can exist because of the many ways that the two can combine, orient, or segregate within structures. Neither moiety possesses an inversion centre. Therefore, when studying current actinide borate chemistry, this should be recognized and learned quickly. In the present example, it is the $BO_4$ tetrahedra which directly chelate the $Th^{4+}$ centre, whereas the $BO_3$ triangles only share vertices with the larger thorium polyhedra. These thorium centres are 12-coordinate and show icosahedral geometry.

This localized geometric progression spires into a much more revealing topography when viewed at a larger scale. NDTB-1 is considered to be a porous supertetrahedral three-dimensional framework and exhibits channels which form along the plane in the cubic environment (Figure 23.1(a)). The channels further combine at the centre of supertetrahedra to form hexagonal cavities (Figure 23.1(b) and (c))— this joint free void space contributed by both the channels and cavities represents 43% of the material's entire volume (Figure 23.1). It was determined that this material might have a significant capability of participating in ion change since the weakly coordinated $H_2BO_2^-$ species can be easily mobilized into vacating the structure due to an equilibrium with its weak acid counterpart, $H_3BO_3$.

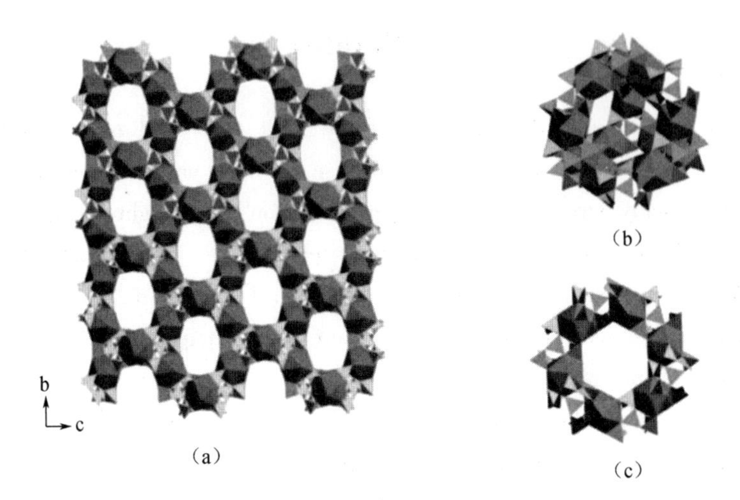


**Figure 23.1 The NDTB-1 supertetrahedral framework channels(a) a cavity(b) and (c) a view of a single hexagonal window from the cavity.**

An outstanding property discovered about this material is that it can participate in what is called single crystal tosingle-crystal ion exchange. This is defined as the ability to engage in ion exchange without exceeding a limiting threshold of perturbations that can disrupt and cause collapse of the crystal structure framework. Most potential ion exchange materials demonstrate collapse because the combination of charge mobility combined with a destabilization of the coordination environment exceeds this limit. NDTB-1, however, resists collapse.

Within the context of functionalized architectures, borates represent an impressive diversity of materials. Some crystallographic properties for these materials are summarized in Table 23.3. Yet to be fully explored, there are still many untried possibilities for modifying some of the topologies presented. For example, Chen's anionic group theory argues that a reduction of $BO_4$ tetrahedra paired with an increase in $BO_3$ triangles, and especially the creation of the six-membered ring $B_3O_6$, can lead to a significant increase in second-harmonic generation. This is due to the typically larger response of π-orbitals in planar groups toward polarized light. If some of these advances can be realized in the actinide series, then it could lead to even more impressive and complex materials in the future.

**Table 23.3 Uranium and thorium borates**

| Compound | Space group | Color | $a_0$(Å) | $b_0$(Å) | $c_0$(Å) | β, α, γ(°) | Volume (Å$^3$) |
|---|---|---|---|---|---|---|---|
| $Ca(UO_2)_2B_2O_6$ | *C*2 | — | 16.512 | 8.169 | 6.582 | 96.97 | 881 |

**Table 23.3**(Continued)

| Compound | Space group | Color | $a_0$(Å) | $b_0$(Å) | $c_0$(Å) | β, α, γ(°) | Volume (Å$^3$) |
|---|---|---|---|---|---|---|---|
| $K_6[UO_2(B_{16}O_{24}(OH)_8)]\cdot 12H_2O$ | $P2_1/n$ | — | 12.024 | 26.45 | 12.543 | 94.74 | 3 975 |
| $[ThB_5O_6(OH)_6][BO(OH)_2]\cdot 2.5H_2O$ | $Fd_3$ | Colorless | 17.403 6 | 17.403 6 | 17.403 6 | 90.0 | 5 271.3 |
| $Li[(UO_2)B_5O_9]\cdot H_2O$ | $Pn$ | Yellow-green | 6.378 3 | 6.224 1 | 10.530 8 | 89.996 | 418.06 |
| $Tl[UO_2)B_5O_8(OH)F]$ | $P1$ | Light yellow | 6.416 3 | 6.466 3 | 7.135 7 | 92.1, 103.3, 119.6 | 246.67 |

# 23.3 Sulfates

### Thorium

The simplest thorium sulfates come in a number of hydrated varieties. The first example to show the generic mode of bonding is from that of the octohydrate, $Th(SO_4)_2\cdot 8H_2O$. The thorium centre has a 10-coordinate bicapped square antiprismatic geometry and coordinates four total oxygens from two sulfate groups with a complement of six waters. It is important to note that the two additional waters do not bind to the thorium centre but instead form bridging hydrogen bonds between additional coordinated thorium polyhedra. Due to the anisotropy of coordination, the $SO_4$ tetrahedra have varying S-O bonds of approximately 0.04 Å difference, with the coordinated bonds being longer than the terminal bonds.

The next example in progression of complexity is that of a ternary thorium sulfate, $Na_2[Th(SO_4)_3(H_2O)_3]\cdot 3H_2O$. The thorium centre in this example shows a reduction in coordination number with a higher chelating number of sulfate groups involved in the bonding. The division of bonds, as scaled in line with the first example, totals six oxygens from bidentate sulfate groups and three equatorial oxygens from inner-sphere coordinating water molecules, making the thorium centre adapt itself to be a nine-coordinate, tricapped trigonal prism. The sulfate groups serve as the bridging moieties between thorium centres, and this extension ultimately creates a chain topography. As before, S-O coordinated bonds are longer than S-O terminal bonds, although the mean difference in bond length is only about 0.03 Å in this example. Furthermore, the equatorial water bond lengths are found to be longer than the bridging sulfate oxygens, which indicates the ease with which such a compound is hydrated.

### Uranium

Uranium sulfates, are divided primarily into those with $UO_2^{2+}$ functional groups versus those with $U^{4+}$ groups. The first basic structure worth observing is that of $U(SO_4)_2\cdot 4H_2O$. One immediate difference from thorium structures to note is that $U^{4+}$ has a smaller ionic radius and thus will exhibit smaller coordination numbers. In the structure at hand, uranium adopts an eight-coordinate distorted square antiprism geometry. Four monodentate sulfate groups coordinate it while bridging to equivalent uranium centres. Because all of the sulfate groups bridge between metal centres, a topography of successive layers becomes the main feature. Water fills the remaining coordination sites and takes up occupational spaces between the layers. Analogous to the presented thorium structures, a significant quantity of hydrogen bonds decorate the overall lattice.

Another important structure is that of $Cs_2U(SO_4)_3 \cdot 2H_2O$, which crystallizes in the monoclinic space group $P2_1/c$. This structure has similar features to its analogue $Cs_2Th(SO_4)_3 \cdot 2H_2O$ in that both coordinate a total of five sulfate groups and two waters. Four of these sulfate groups are bidentate and bridge to additional metal centres to create topographically layered sheets. It is important to note that since uranium has a lower coordination number of nine that the fifth terminal bidentate linkage in the thorium compound becomes a terminal monodentate linkage in the uranium structure.

## 23.4 Phosphates

Actinide phosphates are another class of widely studied structures. One of the largest hopes with the phosphates is that they will be used as a potential host matrix for nuclear waste disposal. The phosphate anion, like borate and sulfate, is characterized by a nonmetallic bond which possesses strong covalent character. In comparison, both the S-O and P-O bonds are more covalent than the B-O bond. However, because $PO_4^{3-}$ holds an extra charge compared to $SO_4^{2-}$, it prefers insoluble phases when paired with the predominantly trivalent and tetravalent actinides. The number and abundance of these phases and structures has made phosphates a popular subject of study with actinides, and only a brief survey of these will be given.

All of the trivalent orthophosphates from plutonium to einsteinium have been made and characterized. It is notable that the stability of these binary compounds can be increased by producing the monazite crystalline phase typically through heating at very high temperatures. The monazite structure, which is also the monoclinic space group $P2_1/c$, can be found in nature and is one the most predominant forms that lanthanides are found to exist in. It also incorporates both tetravalent uranium and thorium into its structure through a variety of substitutive mechanisms.

The most common divalent substitution partner for an actinide is calcium. The ability of monazite to be able to incorporate nearly all trivalent and tetravalent actinides, along with its chemical and radiation stability have made it a promising candidate for a suitable waste form. This is especially true for waste streams generated from reprocessing with molten salt reactors, because the actinide fluorides are much more insoluble and can leach from traditional borosilicate glass waste forms. The monazite structure, as well as other phosphate-based glasses such as sodium aluminophosphate glass, are good alternatives to consider for this. An example of the monazite structure is given in Figure 23. 2 for the recently solved $PuPO_4$ structure which has alluded researchers for some time.

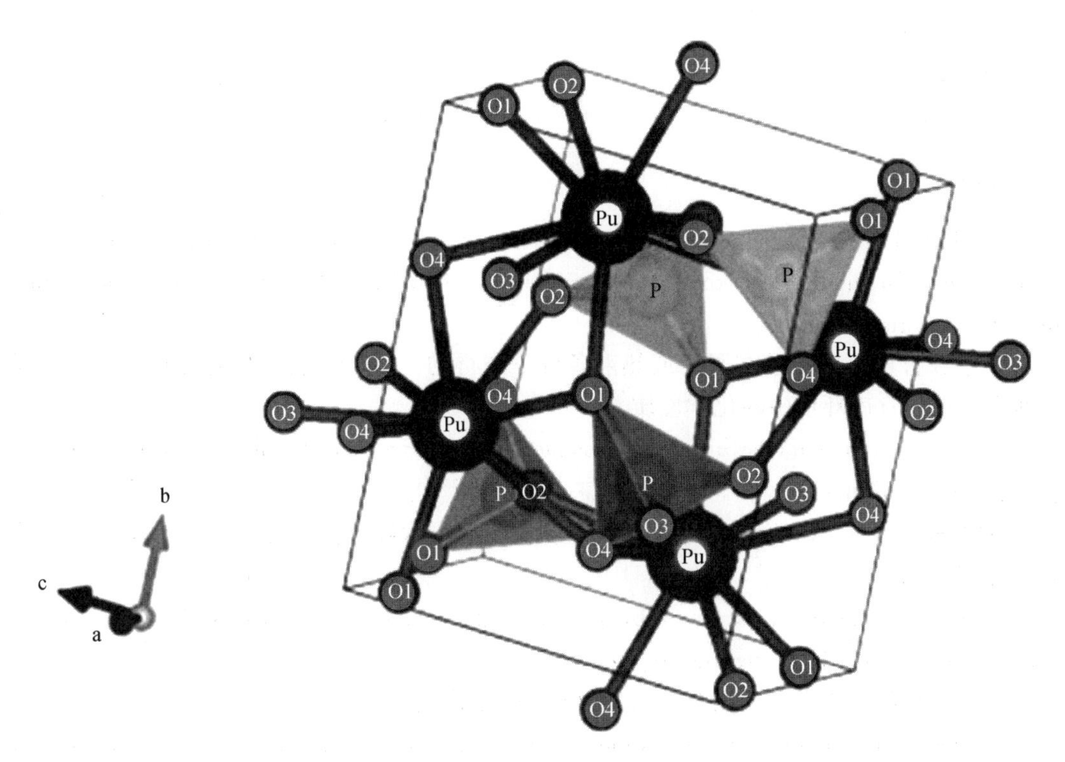


**Figure 23.2 The monazite structure for trivalent $PuPO_4$, which has been difficult to isolate**

## 词汇

fluoride 氟化物
centrifuge 离心机
orthorhombic 正交晶系的
stoichiometric ratio 化学计量比
borate 硼酸盐
crystallize (使)结晶
single-crystal 单晶
topology 拓扑学
crystallographic 晶体学的
tetrahedra 四面体
coordination 配位
anisotropy 各向异性
polyhedral 多面体的
sulfate 硫酸盐
monodentate 单齿配位
bidentate 双齿配位
monoclinic 单斜晶系的

| | |
|---|---|
| phosphate | 磷酸盐 |
| orthophosphate | 正磷酸盐 |
| borosilicate | 硼硅酸盐 |
| monazite | 独居石 |

**注释**

1. Most of the synthetic reactions are conducted in either the solid state or hydrothermally-in general, this also mimics the expectations for chemistry found in either reactor conditions or in a repository environment.

大多数合成反应都是在固态或热液条件下进行的,一般来说,这也是在模拟预期的反应堆条件或储存库环境中的化学性质。

2. The need to characterize fluoride complexes first came from the ubiquitous use of $UF_6$ in gas centrifuges for the isotopic enrichment of $^{235}U$ from $^{238}U$.

表征氟化物络合物的需要,首先来自气体离心机中六氟化铀的普遍使用,以从$^{238}U$中同位素富集$^{235}U$。

3. The trivalent fluorides crystallize predominantly in the hexagonal crystal system and are isostructural with the $LaF_3$ structure type.

三价氟化物主要在六方晶系中结晶,与$LaF_3$结构类型同构。

4. Depending on the actinide, different treatments to achieve the trivalent fluorides are necessary.

根据锕系元素的不同,需要进行不同的处理以获得三价氟化物。

5. The tetravalent fluorides crystallize in the monoclinic crystal system and are isostructural with the $ZrF_4$ structure type.

四价氟化物在单斜晶系中结晶,与$ZrF_4$结构类型同构。

6. Higher temperatures are possible with these designs, which can allow for more efficient burn-up.

这些设计可以实现更高的温度,从而提高燃耗效率。

7. All of the hexavalent fluorides exist in the tetragonal crystal system and have low-boiling temperatures.

所有六价氟化物都存在于四方晶系中,并且具有低沸点温度。

8. Actinide borates are considered to be one of the most successful oxyanions for probing structural variation across the 5f-series.

锕系元素硼酸盐被认为是探测 5f 系列结构变化最成功的含氧阴离子之一。

9. An outstanding property discovered about this material is that it can participatein what is called single crystal to single-crystal ion exchange.

这种材料的一个突出特性是,它可以参与所谓的单晶与单晶的离子交换。

10. One of the largest hopes with the phosphates is that they will be used as a potential host matrix for nuclear waste disposal.

磷酸盐的最大希望之一是,它们将被用作核废料处理的潜在主基质。

11. The ability of monazite to be able to incorporate nearly all trivalent and tetravalent

actinides along with its chemical and radiation stability have made it a promising candidate for a suitable waste form.

独居石有能够掺入几乎所有三价和四价锕系元素的能力,而且具有化学和辐射稳定性,这使它有望成为一种合适的废物形式。

# Chapter 24 Radionuclide Generators

Radionuclide generator systems continue to play a key role in providing both diagnostic and therapeutic radionuclides for various applications in nuclear medicine, oncology and interventional cardiology. Although many parent/daughter pairs have been evaluated as radionuclide generator systems, there are a relatively small number of generators, which are currently in routine clinical and research use. Essentially every conceivable approach has been used for parent/separation strategies, including sublimation, thermochromatographic separation, solvent extraction, and adsorptive column chromatography. The most widely used radionuclide generator for clinical applications is the $^{99}Mo/^{99m}Tc$ generator system, but recent years has seen an enormous increase in the use of generators to provide therapeutic radionuclides, which has paralleled the development of complementary technologies for targeting agents for therapy and in the general increased interest in the use of unsealed therapeutic radioactive sources.

## 24.1 Historical Perspective

A radionuclide generator is a concept defined as an effective radiochemical separation of decaying parent and daughter radionuclides such that the daughter is obtained in a pure radionuclidic and radiochemical form. Radionuclide generators were historically called "cows" since the daughter radioactivity was "milked" (i. e., removed) from its precursor and the parent then generated a fresh supply of the daughter.

Generator parent radionuclides are obtained from uranium fission products (i. e. $^{99}Mo$ and $^{90}Sr$) or as decay products from $^{233}U$ ($^{229}Th/^{225}Ac$), or are produced directly in nuclear reactors ($^{188}W$, etc.) or at accelerators ($^{82}Rb$, $^{62}Zn$, etc.).

Compared to in-house radionuclide production facilities such as accelerators or nuclear reactors, the availability of short-lived radionuclides from radionuclide generators provides an inexpensive and convenient alternative. The development of radionuclide generators over the past 3 decades was primarily motivated by the increasing spectrum of applications of radionuclides and labelled compounds in the life sciences, in particular for diagnostic applications in nuclear medicine. In the past years, however, promising applicants of generator-derived therapeutic radionuclides have been developed in the fields of nuclear medicine, oncology, and interventions cardiology. This increasing importance of radionuclide generators has initiated a broad development for radionuclide production of the generator parent radionuclide, for sophisticated radiochemical separations as well as reliable technical design of the generator systems.

The first generator for life sciences application was developed in 1920, providing $^{222}Rn$ ($T_{1/2}=3.825$ d) to obtain radon seeds for radiation therapy as a daughter of $^{226}Ra$ ($T_{1/2}=1.60\times$

$10^3$ a) (Failla, 1920). However, practical importance of radionuclide generators was achieved in 1951 by the $^{132}Te$ ($T_{1/2}$ = 3.2 d)/$^{132}I$ ($T_{1/2}$ = 1.39 h) generator (Winsche et al., 1951), and, much more, in 1957 by the pioneering development of the $^{99}Mo/^{99m}Tc$ generator at Brookhaven National Laboratory (BNL) (Stang et al., 1954; 1957). The technetium daughter radionuclide was soon envisioned for medical use (1960), and indeed its first clinical application was reported in 1961 (Richards 1960; Harper et al., 1962) and has revolutionized radiopharmaceutical chemistry and nuclear medicine. Since that time, various other generator systems have been developed, and some of them received significant practical applicant.

The broad use of the $^{99}Mo/^{99m}Tc$ generator system in nuclear medicine is a key example which has been crucial for more than 2 decades for the hospital or central radiopharmacy preparation of a wide variety of diagnostic agents for applications in nuclear medicine and oncology. Over 35 000 diagnostic procedures are estimated to be currently conducted daily in the USA(> 16 million studies per year) with $^{99m}Tc$. This reliance on the availability of $^{99m}Tc$ clearly underscores the crucial importance of the continued and reliable $^{99}Mo$ production and processing facilities to ensure the uninterrupted study of the generator patent radionuclide required for fabrication of these generator systems.

## 24.2 Generator-produced Positron Emitters

Among the generator pairs relevant for quantitative PET (Table 24.1), all parent nuclides are neutron deficient and are thus produced at accelerators. All daughter nuclides are positron emitters and provide a significant positron branching, although in some cases the decay is accompanied by high-energy photons, which might require careful adoption of PET scanners.

The generators can be categorized according to the half-life of the daughter nuclide. The short-lived daughters cover half-lives of a few minutes. As the short half-lives do not allow radiochemical synthesis, these systems are relevant for perfusion imaging exclusively. The generator design must allow for direct application of the separated daughter for human use.

The longer-lived daughter nuclides, on the other hand, provide a potential for the development of labelled radiopharmaceuticals. However, due to the long half-life and the low cross sections, in particular for parent nuclides $^{44}Ti$, $^{68}Ga$, and $^{82}Sr$, the production rates are relatively low and require long high-current irradiations. Although this results in rather high cost per generator, the number of PET scans achievable lowers the costs per individual patient investigation.

**Table 24.1 Generator-produced position emitters with potential for position emission tomography(PET)**

| Generator system | Parent | Daughter | | | |
|---|---|---|---|---|---|
| | $T_{1/2}$ | $T_{1/2}$ | $\beta_{branch^+}$/% | $E_{\beta^+}$/MeV | Application |
| $^{82}Sr/^{82}Rb$ | 25.6 d | 1.27 min | 95.0 | 1.41 | Perfusion |

**Table 24.1**(Continued)

| Generator system | Parent | Daughter | | | |
|---|---|---|---|---|---|
| | $T_{1/2}$ | $T_{1/2}$ | $\beta_{branch^+}$/% | $E_{\beta^+}$/MeV | Application |
| $^{140}Nd/^{140}Pr$ | 3.37 d | 3.39 min | 51 | 0.544 | Perfusion |
| $^{118}Te/^{118}Sb$ | 6.00 d | 3.60 min | 74 | 0.882 | Perfusion |
| $^{122}Xe/^{122}I$ | 20.10 h | 3.60 min | 77 | 1.090 | Labeling |
| $^{128}Ba/^{128}Cs$ | 2.43 d | 3.62 min | 69 | 0.869 | Perfusion |
| $^{134}Ce/^{134}La$ | 3.16 d | 6.40 min | 63 | 0.756 | Perfusion |
| $^{62}Zn/^{62}Cu$ | 9.26 h | 9.74 min | 97 | 1.280 | Labeling; Perfusion |
| $^{52}Fe/^{52m}Mn$ | 8.28 d | 21.10 min | 97 | 1.130 | Perfusion |
| $^{68}Ge/^{68}Ga$ | 270.80 d | 1.135 h | 89 | 0.740 | Labeling; Perfusion |
| $^{110}Sn$/110mIn | 4.10 h | 1.15 h | 62 | 0.623 | Labeling |
| $^{44}Ti/^{44}Sc$ | 60.30* a | 3.927 h | 94 | 0.597 | Labeling |
| $^{72}Se/^{72}As$ | 8.40 d | 1.083 d | 88 | 1.020 | Labeling |

* Means value of most recent various literature data.

## 24.3 Generator-produced Photon Emitters

Although the use of ultra short-lived generator-derived radionuclides for first-pass radionuclide angiography (FPRNA) for evaluation of ventricular function (wall motion) is not currently widely practiced, the development and use of generator systems for evaluation of pulmonary ventilation and cardiac function was a major research area in the 1970s and 1980s [see, e.g., Paras and Thiessen (1985)]. High count-rate imaging systems and advanced computer technology are required for data acquisition, storage, and analysis. High levels of activity are required for FPRNA, but the short half-lives permit rapid, repeat studies, since the vascular recirculation time is longer than the radionuclide physical half-life. The short radionuclide physical half-lives also ensure greatly reduced radiation burden to both personnel and patients. Generators that provide ultra short-lived daughter radionuclides ($T_{1/2} < 1-2$ min) for FPRNA include the $^{195m}Hg/^{195m}Au$ ($T_{1/2} = 30.5$ s) and $^{191}Os/^{191m}Ir$ ($T_{1/2} = 4.94$ s) systems. More recently, the $^{178}W/^{178}Ta$ ($T_{1/2} = 9.31$ min) generator has been introduced. For pulmonary ventilation studies and evaluation of the right ventricular chamber, the $^{81}Rb/^{81m}Kr$ generator is commercially available and approved for human use in Europe.

Several other generator systems have been developed providing photon-emitting daughter nuclides, but did not receive adequate medical attention. Others were proposed for basic radiochemical and radiopharmaceutical studies rather than for a direct application of the daughter nuclide in nuclear medicine.

$^{77}$Br ($T_{1/2}$=2.377 d)/$^{77m}$Se ($T_{1/2}$=17.4 s). $^{77m}$Se decays by isometric transition with the emission of 162 keV photons in 50% abundance and appeared ideal for radionuclide angiocardiography and to study the effects of exercise and pharmaceuticals on hemodynamics (Lambrecht et al., 1977; Norton et al., 1978; Madhusudhan et al., 1979).

$^{109}$Cd ($T_{1/2}$=1.267 a)/$^{109m}$Ag ($T_{1/2}$=39.6 s). This generator was suggested for FPRNA although the 88 keV photon emission of $^{109m}$Ag shows a rather low abundance of 3.73 % of the $^{109}$Cd decay (Steinkruger et al., 1986). While Bartos and Bilewicz (1995) used crystalline antimony acid as absorbent yielding increasing $^{109}$Cd breakthrough, Mansur et al. (1995) used cation exchange chromatography, but needed a scavenger column to reduce the $^{109}$Cd breakthrough of $10^{-4}$%.

$^{113}$Sn ($T_{1/2}$=115.09 d)/$^{113m}$In ($T_{1/2}$=1.658 h). Because of its long shelf life, the $^{113}$Sn/$^{113m}$In generator has been proposed for labeling and evaluation of various radiopharmaceuticals with the 392 keV photon-emitting $^{113m}$In as an analogue of $^{111}$In (Liu et al., 1989). $^{113}$Sn production (Qaim and Döhler, 1984) and generators were described (Seidl and Lieser, 1973; Rao et al., 1976, 1977; Camin, 1977; Al-Janabi and Al-Hashini, 1979; Lin et al., 1982).

$^{115}$Cd ($T_{1/2}$=2.228 d)/$^{115m}$In ($T_{1/2}$=4.486 h). This generator has been considered as an alternative to the $^{113}$Sn/$^{113m}$In system. Parent radionuclide production and generator design (Ramamoorthy and Mani, 1976; Bhattacharya and Basu, 1979; Erhardt et al., 1981; Yagi et al., 1982) were investigated.

## 24.4 Generator-produced Particle Emitters for Therapy

During the last decade there has been a tremendous increase in the development and use of new therapeutic radiopharmaceuticals radiolabeled with radionuclides, which are available from radionuclide generator systems. The availability of generator-derived therapeutic radionuclides is necessary for the development and testing and commercialization of agents with potential for endoradiotherapy (ERT). Just as availability of $^{99m}$Tc from the $^{99}$Mo/$^{99m}$Tc has played such a key role in the development of a wide variety of $^{99m}$Tc-labelled radiopharmaceuticals, the availability of generator-derived radionuclides has stimulated the development of an increasing spectrum of therapeutic tracers. It is important to note that in most cases, increased need for and further development of generators, which provide therapeutic radionuclides has been driven by the success in the development of the targeting agents or vectors.

Generator-derived therapeutic radionuclides have a number of characteristic decay processes, and can emit β particles, Auger electrons, low-energy photons, and α particles. Since many therapeutic radionuclides are characterized by β decay, they are often directly produced in a nuclear reactor, since neutron capture by target nuclide forms a radioactive or unstable product that decays by β emission. Key examples of therapeutic radionuclides obtained from reactor-produced parent radionuclides include $^{166}$Ho from the $^{166}$Dy/$^{166}$Ho generator, and $^{188}$Re from the $^{188}$W/$^{188}$Re generator.

Another important source of generator parent radionuclides is recovery of generator parent radionuclides that are produced during nuclear fission. Strontium-90 is the parent for the $^{90}Sr/^{90}Y$ generator system and is isolated from fission products. The recent approval by the US Food and Drug Administration (FDA) on February 19, 2002, for "Zevalin" (Ibritumomab tiuxetan) — a $^{90}Y$-labelled murine anti CD20 antibody — for the treatment of patients with low grade, follicular, or transformed non-Hodgkin's lymphoma, represents the first antibody radiolabeled with a therapeutic, generator-derived radionuclide, and world be expected to represent the first of many new therapeutic radiopharmaceuticals for oncologic applications.

A third source for radionuclide generator systems is the recovery of radioactive parents from "extinct" radioactive decay processes, such as $^{229}Th$, which is recovered from $^{233}U$ decay products. The $^{229}Th$ represents a convenient, long-lived ($T_{1/2}$ = 7 340 a) source from which $^{225}Ac$ is recovered, which is the parent of the $^{225}Ac/^{213}Bi$ generator system.

### 词汇

| | |
|---|---|
| therapeutic | 治疗的 |
| oncology | 肿瘤学 |
| cardiology | 心脏病学 |
| chromatography | 色谱分析法 |
| accelerator | 加速器 |
| parent nuclide | 母体核素 |
| angiography | 血管造影术 |
| pulmonary | 肺的 |
| vascular recirculation | 血管循环 |
| tremendous | 巨大的 |
| antibody | 抗体 |
| lymphoma | 淋巴瘤 |

### 注释

1. Radionuclide generator systems continue to play a key role in providing both diagnostic and therapeutic radionuclides for various applications in nuclear medicine, oncology and interventional cardiology.

放射性核素发生器系统在为核医学、肿瘤学和介入心脏病学的各种应用提供诊断和治疗放射性核素方面,继续发挥着关键作用。

2. A radionuclide generator is a concept defined as an effective radiochemical separation of decaying parent and daughter radionuclides such that the daughter is obtained in a pure radionuclidic and radiochemical form.

放射性核素发生器是一个概念,定义为衰变的母体和子体放射性核素的有效放射化学分离,从而以纯放射性核素和放射化学形式获得子体。

3. Compared to in-house radionuclide production facilities such as accelerators or nuclear reactors, the availability of short-lived radionuclides from radionuclide generators provides an

inexpensive and convenient alternative.

与加速器或核反应堆等内部放射性核素生产设施相比,放射性核素发生器所提供的短寿命放射性核素是一种廉价且方便的替代方案。

4. This increasing importance of radionuclide generators has initiated a broad development for radionuclide production of the generator parent radionuclide, for sophisticated radiochemical separations as well as reliable technical design of the generator systems.

放射性核素发生器的重要性日益增加,促进发生器母体放射性核素的放射性核素生产、复杂的放射化学分离以及发生器系统的可靠技术设计的广泛发展。

5. All daughter nuclides are positron emitters and provide a significant positron branching, although in some cases the decay is accompanied by high-energy photons, which might require careful adoption of PET scanners.

所有子核素都是正电子发射体,并提供大量正电子分支,但在某些情况下,衰变伴随着高能光子,这可能要求谨慎采用 PET 扫描仪。

6. The generators can be categorized according to the half-life of the daughter nuclide.

发生器可以根据子核素的半衰期进行分类。

7. The longer-lived daughter nuclides, on the other hand, provide a potential for the development of labelled radiopharmaceuticals.

另一方面,寿命较长的子核素为标记放射性药物的开发提供了潜力。

8. The short radionuclide physical half-lives also ensure greatly reduced radiation burden to both personnel and patients.

短的放射性核素物理半衰期也确保大大减轻人员和患者的辐射负担。

9. Generator-derived therapeutic radionuclides have a number of characteristic decay processes, and can emit β particles, Auger electrons, low-energy photons, and α particles.

发生器衍生的治疗性放射性核素具有许多特征衰变过程,可以发射 β 粒子、俄歇电子、低能光子和 α 粒子。

10. Another important source of generator parent radionuclides is recovery of generator parent radionuclides that are produced during nuclear fission.

发生器母体放射性核素的另一个重要来源是,回收核裂变过程中产生的发生器母体放射性核素。